Studies in Systems, Decision and Control

Volume 382

Series Editor

Janusz Kacprzyk, Systems Research Institute, Polish Academy of Sciences, Warsaw, Poland

The series "Studies in Systems, Decision and Control" (SSDC) covers both new developments and advances, as well as the state of the art, in the various areas of broadly perceived systems, decision making and control–quickly, up to date and with a high quality. The intent is to cover the theory, applications, and perspectives on the state of the art and future developments relevant to systems, decision making, control, complex processes and related areas, as embedded in the fields of engineering, computer science, physics, economics, social and life sciences, as well as the paradigms and methodologies behind them. The series contains monographs, textbooks, lecture notes and edited volumes in systems, decision making and control spanning the areas of Cyber-Physical Systems, Autonomous Systems, Sensor Networks, Control Systems, Energy Systems, Automotive Systems, Biological Systems, Vehicular Networking and Connected Vehicles, Aerospace Systems, Automation, Manufacturing, Smart Grids, Nonlinear Systems, Power Systems, Robotics, Social Systems, Economic Systems and other. Of particular value to both the contributors and the readership are the short publication timeframe and the world-wide distribution and exposure which enable both a wide and rapid dissemination of research output.

Indexed by SCOPUS, DBLP, WTI Frankfurt eG, zbMATH, SCImago.

All books published in the series are submitted for consideration in Web of Science.

More information about this series at http://www.springer.com/series/13304

Samsul Ariffin Abdul Karim

Editor

Shifting Economic, Financial and Banking Paradigm

New Systems to Encounter COVID-19

 Springer

Editor
Samsul Ariffin Abdul Karim 🆔
Fundamental and Applied Sciences Department
and Centre for Systems Engineering (CSE)
Institute of Autonomous System
Universiti Teknologi PETRONAS, Seri Iskandar,
Perak Darul Ridzuan, Malaysia

ISSN 2198-4182   ISSN 2198-4190   (electronic)
Studies in Systems, Decision and Control
ISBN 978-3-030-79612-9   ISBN 978-3-030-79610-5   (eBook)
https://doi.org/10.1007/978-3-030-79610-5

This Springer imprint is published by the registered company Springer Nature Switzerland AG
The registered company address is: Gewerbestrasse 11, 6330 Cham, Switzerland

Preface

This book is a collection of the works that have been conducting by researchers at Universiti Teknologi PETRONAS (UTP), FPT University, Universiti Putra Malaysia (UPM), Universiti Kebangsaan Malaysia (UKM), and Universiti Malaysia Sarawak (UNIMAS). The main idea is started based on the collaboration between UTP and FPT University, Vietnam. In this book, all the authors tackle the main ideas on shifting the economy, finance, and banking sectors among ASEAN counties into a new paradigm. The researchers are using econometric, mathematics, statistics, and quantitative sciences to study many economic, finance, and banking events such as cryptocurrency, consumer preferences, and good governance. Various new and novel results, methods, and algorithms are presented in detail that will benefit the ASEAN policymakers and relevant agencies. The editor would like to express their gratitude to all the contributing authors for their great efforts and full dedication in preparing the manuscripts for the book. We would like to thank all reviewers for reviewing all manuscripts and providing very constructive feedback. The first editor is fully supported by Universiti Teknologi PETRONAS (UTP) and the Ministry of Education, Malaysia through a research grant **FRGS/1/2018/STG06/UTP/03/1015MA0-020(New rational quartic spline for image refinement)** and **YUTP: 015LC0-315 (Uncertainty estimation based on Quasi-Newton methods for Full Waveform Inversion (FWI)).**

This book is suitable for postgraduate students, researchers as well as other scientists who working in econometric, finance, banking, and numerical simulation. Any feedback can be directed to the first editor.

Seri Iskandar, Malaysia
July 2021

Samsul Ariffin Abdul Karim

v

Contents

Editors and Contributors

About the Editors

Samsul Ariffin Abdul Karim is a senior lecturer at Fundamental and Applied Sciences Department, Universiti Teknologi PETRONAS (UTP), Malaysia. He has been in the department for more than 12 years. He obtained his B.App.Sc., M.Sc., and Ph.D. in Computational Mathematics & Computer Aided Geometric Design (CAGD) from Universiti Sains Malaysia (USM). He is a Professional Technologists registered with Malaysia Board of Technologists (MBOT). He had 20 years of experience using Mathematica and MATLAB software for teaching and research activities. His research interests include curves and surfaces designing, geometric modeling, and wavelets applications in image compression and statistics. He has published more than 140 papers in Journal and Conferences as well as eight books including two research monographs and three Edited Conferences Volume and 40 book chapters. He was the recipient of Effective Education Delivery Award and Publication Award (Journal & Conference Paper), UTP Quality Day 2010, 2011, and 2012, respectively. He was Certified WOLFRAM Technology Associate, Mathematica Student Level. He has published six books with Springer.

Contributors

Aisyah Abdul Rahman Faculty of Economics and Management, Universiti Kebangsaan Malaysia (UKM), Bangi, Selangor, Malaysia

G. Venkata Ajay Kumar Kadapa, India

Soan T. M. Duong Le Quy Don Technical University, Hanoi, Vietnam

Hanny Zurina Hamzah School of Business and Economics, Universiti Putra Malaysia, Serdang, Malaysia

Yi-Chung Hu Department of Business Administration, Chung Yuan Christian University, Taoyuan City, Taiwan

Suryati Ishak School of Business and Economics, Faculty of Economics and Management, Universiti Putra Malaysia, Serdang Selangor, Malaysia

Amin Jan Faculty of Hospitality, Tourism, and Wellness, Universiti Malaysia Kelantan, Pengkalan Chepa, Malaysia

Bakri Abdul Karim Faculty of Economics and Business, Universiti Malaysia Sarawak (Malaysia), Kota Samarahan, Sarawak, Malaysia

Samsul Ariffin Abdul Karim Fundamental and Applied Sciences Department and Centre for Systems Engineering (CSE), Institute of Autonomous System, Universiti Teknologi PETRONAS, Seri Iskandar, Perak Darul Ridzuan, Malaysia

Zulkefly Abdul Karim Center for Sustainable and Inclusive Development (SID), Faculty of Economics and Management, The National University of Malaysia (UKM), Bangi, Selangor, Malaysia

Norlin Khalid Faculty of Economics and Management, Universiti Kebangsaan Malaysia (UKM), Bangi, Selangor, Malaysia

Maran Marimuthu Department of Management and Humanities, Universiti Teknologi PETRONAS, Seri Iskandar, Perak Darul Ridzuan, Malaysia

Mehreen Mehreen Department of Management and Humanities, Universiti Teknologi PETRONAS, Seri Iskandar, Perak Darul Ridzuan, Malaysia

Syajarul Imna Mohd Amin Faculty of Economics and Management, Universiti Kebangsaan Malaysia (UKM), Bangi, Selangor, Malaysia

Sivaramakrishnan Natesan Department of Mechanical Engineering, Saveetha Engineering College, Chennai, India

Lan T. M. Nguyen FPT University, Hanoi, Vietnam

Phi-Hung Nguyen Department of Business Management, National Taipei University of Technology, Taipei, Taiwan;
Faculty of Business, FPT University, Hanoi, Vietnam

Ahmad Khidir Othman Center for Sustainable and Inclusive Development (SID), Faculty of Economics and Management, The National University of Malaysia (UKM), Bangi, Selangor, Malaysia

N. G. Praveena Department of Electronics and Communication Engineering, R.M.K College of Engineering and Technology, Chennai, India

R. R. Rajalaxmi Department of Computer Science and Engineering, Kongu Engineering College, Perundurai, India

Sharul Shahida Shakrein Safian Department of Economics and Financial Studies, Faculty of Business Magement, Universiti Teknologi Mara (UiTM), Selangor, Malaysia

K. Tamilarasi Department of Computer Science and Engineering, Jeppiaar Institute of Technology, Sriperumpudur, India

Nguyen Thi Lieu Trang Department of International Business, FPT University, Hanoi, Vietnam

Jung-Fa Tsai Department of Business Management, National Taipei University of Technology, Taipei, Taiwan

Chockalingam Aravind Vaithilingam High Impact Research Laboratories, Faculty of Innovation and Technology, Taylor's University, Subang Jaya, Malaysia

Toh Kim Yuan School of Business and Economics, Faculty of Economics and Management, Universiti Putra Malaysia, Serdang Selangor, Malaysia

Mohd Azlan Shah Zaidi Center for Sustainable and Inclusive Development (SID), Faculty of Economics and Management, The National University of Malaysia (UKM), Bangi, Selangor, Malaysia

Evolution of Outbreaks, Lessons Learnt and Challenges Towards "New Normalcy"—Post COVID-19 World

Chockalingam Aravind Vaithilingam, Sivaramakrishnan Natesan,
R. R. Rajalaxmi, K. Tamilarasi, N. G. Praveena,
and Samsul Ariffin Abdul Karim ⓘ

Abstract A new era in epidemics started due to unhealthy practices, population density, environmental changes, migration and deforestation. The rapidity in the spread is primarily due to globalization as we moved to the industrial revolution where everything is internet-connected. In past 30 years, the trend exhibits an increase in the number of epidemics challenging the social well-being, the economy and to some extent the national security. And this translates to the impact on the industrial growth, the race of future together fighting with the newest of the viruses. This paper analyzes and reviews the outbreaks from the start of the revolutionary steam power generation to the modern days, their impact to generate new values to the society

C. A. Vaithilingam (✉)
High Impact Research Laboratories, Faculty of Innovation and Technology, Taylor's University, 47500 Subang Jaya, Malaysia
e-mail: aravindcv@ieee.org

S. Natesan
Department of Mechanical Engineering, Saveetha Engineering College, Chennai 602105, India
e-mail: nsivaramakrishnan@saveetha.ac.in

R. R. Rajalaxmi
Department of Computer Science and Engineering, Kongu Engineering College, Perundurai 638060, India
e-mail: rrr@kongu.ac.in

K. Tamilarasi
Department of Computer Science and Engineering, Jeppiaar Institute of Technology, Sriperumpudur 631604, India
e-mail: tamilarasik@jeppiaarinstitute.org

N. G. Praveena
Department of Electronics and Communication Engineering, R.M.K College of Engineering and Technology, Chennai 601206, India
e-mail: praveena.ng@rmkcet.ac.in

S. A. Abdul Karim
Fundamental and Applied Sciences Department and Centre for Systems Engineering (CSE), Institute of Autonomous System, Universiti Teknologi PETRONAS, Bandar Seri Iskandar, 32610 Seri Iskandar, Perak Darul Ridzuan, Malaysia
e-mail: samsul_ariffin@utp.edu.my

© Institute of Technology PETRONAS Sdn Bhd 2022
S. A. Abdul Karim (eds.), *Shifting Economic, Financial and Banking Paradigm*,
Studies in Systems, Decision and Control 382,
https://doi.org/10.1007/978-3-030-79610-5_1

1

that translated the newer solutions to become new norms. Impact on the outbreaks on the various key sectors and the measures that lead us to overcome is presented. We present the new normal which would become normal in the near future, the post COVID-19 scenario.

Keywords Corona virus · New normalcy · COVID-19 · Innovations · IR 4.0 · Outbreaks

1 Introduction

The global spread of a number of pandemics are on the rise during recent years more to blame globalization. The spread of civilizations, exploration and expanding of trade routes over centuries bring newer types of virus and bacterial infection primarily through human to human transmission or through a host living cell. Table 1 draws a comprehensive role on the various outbreaks and categorized towards the industrial revolution approach. Within the known pandemic, smallpox was widely spreading during the early eighteenth century even before the classification of the industrial revolution was in place, with the trade, colonization and occupation of the territory being the most common reason [1, 2].

Vaccination was the newer way to overcome such an outbreak and it is the start of bio-industry prominently start to work on dealing with infectious diseases including cholera, early dengue lately for Influenza-like Illness (ILI) and Severe Acute Respiratory Influenza (SARI) strained diseases. Regional or endemic and epidemic prevailed during the industrial revolution 2.0 and early industrial revolution 3.0 with Human Immunodeficiency Virus HIV-AIDS with the parental source through sexual infection prevail a larger threat [3].

The disease aggressiveness is primarily due to the personal lifestyle of being in a closest as sugar estates, mining places with less knowledge of sexual protections. A number of new types of protective device both for male and female industry dominate the health industry to resist the viral disease. It is estimated close to 5 million people died of HIV infection to date [4].

Parallel government institutions started to create awareness programme and made it compulsory to educate the people when they enter adulthood. The end of industrial revolution 3.0 saw the mutated version of the SARS virus, the host primarily from infected bats. The genome sequence of the virus stain keeps mutating as it moves through different host primarily due to the global movements of people and interests towards exotic meats.

Table 1 Major outbreak towards newer way of life

Industrial Revolution [5]	Type	Known Origin	Estimated Mortality Rate	Known Source	Change over	Lesson Learned	Industry affected	Newer Solutions	References
Known civilization happens when energy is needed for basic needs									
1	Smallpox (15th–17th centuries)	North America	3 out of 10	Trace on Mummies	Currency against gold	Changed how money was valued towards modern capitalism	Health and Livability	First known Vaccination	[1, 6–9]
							Cotton		
	Cholera pandemic (1817–1876)	India	Still prevail	Water	Cleanliness	Importance of proper, modern sanitation towards the wealth inequality amongst countries	Health and Livability,	Sanitation	[10–12]
							Trade and Textiles		
Localized disease originated prevail to move due to trade between countries and WWI									
2	Spanish Flu (1918–1919)	France	25–30% of world population	Bird	Pandemic Research	Stressed the importance of research into outbreaks of massive scales	Trade between countries	Research on Outbreaks	[13–15]
	Hemorrhagic fever (Dengue) (1950)	Philippines, Thailand	Still prevail	Mosquito	Insurance	Household burden making insurance to be handy	Health, Tourism	Mass disinfection	[16, 17]
	H3N2 (1968–1970)	Hong Kong	1 Million globally	Bird	NA	Towards containment of the disease	Tourism	Swine Vaccine	[18]

(continued)

Table 1 (continued)

Start of globalization is viewed towards the latter half making endemic disease outbreak towards global pandemic

Industrial Revolution [5]		Type	Known Origin	Estimated Mortality Rate	Known Source	Change over	Lesson Learned	Industry affected	Newer Solutions	References
3	Partial automation using computer and memory devices	Ebola 1976	Congo	88%	Bat	Societal	Contact tracing	Local tourism	Quarantine	[16, 19]
		HIV AIDS	West Africa	Fewer by 5 Million	Chimpanzee	Societal	Continues to impact the lives of people today and its negative influence social impact	Societal Impacts, LGBT community prevalent	Protective measures	[4]
		Nipah 1999	Malaysia	40–70%	Pig	Poultry	Unusual illnesses in animals and humans, that relate to scientist pattern trace from other virus	pig-farming industry	Pork Industry cull handling	[20–22]
		SARS (2002–2003)	China	14–15%	Pangolin/Bat	Public Sanitation	Regular sanitized and face masks have become a common sight on the street. (COVID-19 play a similar role in countries today).	Several impact in almost every industry	Early detection of ILI and SARS	[23, 24]

(continued)

Table 1 (continued)

Industrial Revolution [5]	Type	Known Origin	Estimated Mortality Rate	Known Source	Change over	Lesson Learned	Industry affected	Newer Solutions	References
Strong globalization and frequent travel and contact between human to human, make way for a number of epidemics and endemics make the outbreak localized, regional and global pandemics									
4 Application of information and communication technologies to industry	Swine Flu (2009 –2010)	US/Mexico	11–12% of global population	Bird	New types of influenza	Persistent vulnerability towards ILI and SARI.	Tourism and Travel	infection control and hygiene, Masks	[25, 26]
	H7N9 Avian Flu 2009	China	40%	Bird	Poultry	Towards containment of the disease	Bird industry	Avian Industry cull handling	[27, 28]
	MERS COV 2012	Saudi Arabia	34.40%	Camel	Contact isolation	Index case and super spreader	Oil trade, Tourism	Contact tracing, PPE, N95 Masks	[29, 30]
	Ebola 2014	Africa	88%	Bat	African economy	Zoonotic like from animal species to humans (as the coronavirus appears to have done).	Regional Economy	Severe Quarantine	[31–35]
	MERS COV 2015	Korea	40%	Human to human	Isolation	Index case and Super spreader	Oil trade, Tourism	Contact tracing, PPE, N95 Masks	[36–38]
	Zika 2015	Brazil	10.50%	Mosquito	Isolation	Mosquito repellant/ disinfection of communities, water stagnation	Local Tourism	Severe Quarantine	[27, 28]

(continued)

Table 1 (continued)

Industrial Revolution [5]	Type	Known Origin	Estimated Mortality Rate	Known Source	Change over	Lesson Learned	Industry affected	Newer Solutions	References
	Yellow Fever 2016	Angola	20–50%	Mosquito	Community Congestion	Mosquito repellant/ disinfection of communities, water stagnation	Local Tourism	Community Isolation	[16, 39, 40]
	SARS CoV-2 (2019–present)	China	3.4% * of world population	Pangolin/Bat	Commerce	Now that the coronavirus has created a socially distanced world. Social distancing, remote working (Zoonotic)	Impact on every other industry	Social distancing, hygiene, Contact tracing	[41–47]

ILI Influenza Like Illness; *SARI* Severe Acute Respiratory Influenza; *WWI* World War I; *PPE* Personal Protection Equipment; *N95* Not resistant oil respirator that blocks 95% of 0.3 micron particles

2 Global Evolution of Outbreaks

The outbreak of epidemics at a certain point of time naturally affects the different sectors of the industry. Thus, the industrial revolution must face unforeseen challenges because of the epidemics. The early stages of swine flu affect the global economy with impacts towards localized tourism [25]. However, it opens the avenue for importance on the anti-viral measures and oppurtunities for outbreak-related industries like mask manufacturing and infection control medicines. Later, the first case of the Middle East Respiratory Syndrome (MERS), was reported in Saudi Arabia in 2012 [36] followed by South Korea with a greater number of cases in mid of June 2015 [29].

The lessons learned from the spread are the importance of early identification, the root cause of the spread of the virus, pandemic nature and assessment of health system. The outbreak of Ebola virus in South Africa [48] demands the need for preparedness, handling risks, public crisis, food security, community engagement, social protection for women and socio-economic issues. The spread severely affects the agriculture sector in West Africa than other sectors [31]. Early diagnosis of the virus would be helpful for the patient and public health administrators. Due to indistinctness in the viral profile, it remains a challenge for the diagnosis in the first few days of the illness.

Even with a highly developed information society, viruses such as SARS-CoV [44], MERS-CoV [29, 38, 44, 48, 49], H5N1 [7–10], H7N9 [7–11], Ebola [19, 31–35], emerging 2019-nCoV [40, 44, 47, 49–56] and other invisible viruses can still cause devastating effects on human beings. Even today dengue is more life-threatening than the other known ones [2]. The case of the biggest pandemic Coronavirus Infectious Disease (COVID-19) a.k.a. novel coronavirus (nCOVID), a recent stain of the SARS CoV2 stain primarily believed to be originated from a wet market in Wuhan, China [44]. It is believed the host transmission could be through the bat-pangolin-human transmission and further through human–human transmission [44, 47]. Figure 1 shows the era of industrial revolution and its mapping to the illness spread over.

A number of parallels are established of this pandemic to the 1918 Spanish flu [13, 14, 57, 58] primarily due to the rapidness at which the virus moves through the host and mutating inside the living cell to grow, to leave less unknown symptom until the respiratory track is more than half infected. In comparison with the 1918 flu that swept the world inevitably there are several parallels and so as the differences with that of the new COVID-19. The spread of both pandemics are similar, with COVID-19 is more rapid across the globe due to globalization while the Spanish flu took just eight weeks to spread in the entire United States.

In terms of the mortality between those Spanish flu, causes a higher ratio for aged between 20 and 40 but as on the day we write SARS CoV2 [59, 60] is more vulnerable to the elderly people, less so to young adults and much less to children. Because of the rapid mutations, these newer types and versions of the pandemic cannot be controlled to major scales. Only thing is to come up with norms for social,

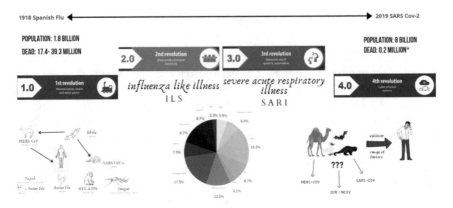

Fig. 1 Industrial revolution outline and the outbreaks

business and governance models that would help to tackle any type of pandemic in the future, the biggest lesson COVID-19 has taught us [47].

Figure 2 shows the footfall of the international tourists between the years 2000 and 2020 until the COVID outbreak [61]. It has been observed that the downfall of tourists up to 3 million which is 0.4% due to the SARS epidemic in the year of 2003. It has been noticed that the downfall for 37 million people which is 4% in the year 2009 due to the global economic crisis. The world tourism organization (UNTWO) estimated that the downfall becomes 290–440 million people which is 20–30% less compared to the post COVID outbreak footfall.

Table 2 shows the potential impact on various industries with possible recommendations to overcome the challenges during and after lockdown. The pandemic is expected to have a huge impact on the global education and skill sector. Most of the

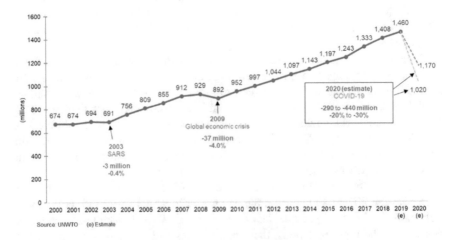

Fig. 2 International tourist arrivals around the world during 2000–2020 [61]

Table 2 Major industry sector post COVID-19

Industry	Sector	Adverse impact	Challenges	Newer approach	References
Auto	Indian automobile manufacturer	Negative impact of revenue at least $1.5–2.0 billions per month across the automobile industry			
		Loss of 7.5 lakhs units in production, unavailability of BS VI models	Balancing the disruption in the supply chain, delay in availability of certified vehicles	Prioritize the supply chain capacity, cash and liquidity management	[64]
	Chinese auto industry	More than 80 percent of the world's auto supply chain is connected to China			
		Car sales plunged by 18%, impact on global automakers due to the production shortfalls in China	To maintain higher production rate after lockdown	Digitizing the sales processes to enhance online sales	[65]
	Electric vehicle	EV's are more expensive than the gas powered vehicle which lead to further decrease in demand of EV's			
		Increase in the payback time	Decrease in demand of EV's due to the excess fossil fuel demand	Apply more stringent emission rules	[66]
Education	Schools	Severe disruption children's social life, learning and skill growth and moved on to ONLINE classes			
		Home schooling with a lot of trial and error in assessment	Cover up the contact hours	Online courses, Webinar	[62, 63]
	Prof. education	Careers and recruitment, availability of skilled labor	Lower paid jobs, unemployment	e-Training delivery	
Construction	Real estate	Alternative investment fund with a total corpus of US$ 3.750 million to bail out 1600 real estate projects to stalled to an acute liquidity crunch			

(continued)

Table 2 (continued)

Industry	Sector	Adverse impact	Challenges	Newer approach	References
		Price variation of key raw material, cash flow, labor force	Turmoil in the equity investment	Relaxation in the project delays	[62]
Financial services	Banking and NBFC	Reduced offtake agreement, fall in banking income	Repayment on capital marketing	Forbearance to service bond, EMI	
Tourism	Tourism and travel	Loss of 300–450 US$ billion in international tourism receipts			
		Travel bans & border closures, cash flow constraints	Losing of jobs	To continue the expansion on tourism despite occasional shocks	[61]
Agriculture	Indian agriculture	People living on agriculture and allied activities, mostly those losing their income			
		Depletion in the food exports, unavailability on food grains, fruits and vegetables, delay in harvesting	Distribution of the commodities	Moratorium on crop loans interest	[67]
	Indian seed corporation	India needs about 250 lakh quintals of seeds for the kharif season			
		Affect the preparation of seeds which generally happens between March and May	Requirement of allied sectors such as transport, testing labs and the packaging industry	Start transporting the farm inputs ie., seeds, grains	[68, 69]

governments around the world have temporarily closed the educational institutions to suppress the spread of COVID-19. The COVID outbreak may have a negative impact on the admissions for the upcoming academic year in the Higher Educational Institutes [62, 63]. Income from tourism has already crippled due to the unprecedented global health emergency.

3   Orbit of Shift Innovation

Many businesses and service industries have restructured themselves quickly by transforming the available resources to cater to the needs of the general public at these unprecedented times. One of the key mantras succeeding in businesses is to understand the present market demand and to usher their business in the right direction. The idea of train compartments being converted to accommodate the COVID-19 infected patients [70], multi-million-dollar cosmetic industries have started manufacturing protective masks [71] and hand sanitizers [72] that are in great demand in the present scenario. Designing life-supporting ventilators in this very short period is the need of the hour, many top-ranking universities and research organizations are actively involved in this area of technological research [73]. Table 3 shows the business flap for a short term with societal or commercial interests during the COVID-19 time to sustain the business. Number of industry players worked with newer collaborators so the shift in business is quiet fast addressing to the demand arise of the pandemic [69, 74–76]. Whatever the industry player the shift in business aligned towards the common product aligned to protective equipment's [71] and usage of drone for various applications [77]. In a way the shift innovation is the biggest outcome/ passive effect for the contemporary players to work with newer players [78].

Table 4 shows the Industry shift in business with a new partner during the COVID-19 Scenario. A number of players shift towards development of the protective equipment's including sanitizers, face shields[73, 79, 80], ventilators and in some players form the engineering industry towards use of drone copters for applications ranging from disinfections, food delivery etc. The primary may be due to the societal support but on the other side this create opportunity to relook the parallel supply making the industry to keep on the roll.

4   Towards the New Normalcy

The number of lessons or the block of pandemic over years has informed us for a new normal always with the first outbreak made us to invest on the virus research [41]. This lead to a number of precautionary vaccines being in place right from the day baby is born to come to adulthood. The immunity level has raised in well since a century below primarily is because of the new normalcy taught us after the first outbreak under the first industrial revolution [41]. The second range of outbreak dominate to make sanitation both personal and also in public become the new normal. The third industrial revolution era predominately taught us to be well aware of the protection primarily force us towards more organized living in private making harmonic living. The current pandemic force us to be well away from each other through the new normal of "social distancing" within the community [47].

Table 3 Collaboration towards sustaining the business during COVID-19 scenario

Industry	Sector	Primary	New collaborator	Shift in business	References
Johnson and Johnson	Beauty	Medical devices, pharma	BARD	Vaccines and immunity medicines	[81]
Ethicon	Hospital	Surgical/wound closure devices	Prisma health	Ventilators	
New balance	Foot wear	Footwear	–	Protective masks	[79]
Ford	Auto	Ford cars, trucks, SUV, EV and luxury vehicles	3M/UAW GE-Fisher Scientific Mahindra & Mahindra	PPE ventilators full face shields	[71]
General motor	Auto	Car makers	Ventec life systems UPS	Protective face shields/gowns	[82]
Tesla motor	Auto	Electric car	Medtronic	Ventilators	[83]
Dyson		Digital motors	Babcock and ventilator challenge UK	Ventilators	[84]
Virgin orbit	Space	Launch of small satellite	Univ. of California, Irvine/Texas, Austin	Bridge ventilators	[85]
ISSINOVA	Design	3D print valves	Sportswear company (Decathlon)	Ventilator	[85]
Uber India	Transport	Travel	Flipkart	Delivers essential to the customer doorstep	[85]
Fedex	E-commerce	Transportation	WING/WALGREENS	Medic delivery drones	[85]
DHL	Logistics	Courier, parcel, and express mail service	*Wingcopter*	Parcelcopter	

When travel by any mode becomes unavoidable the current pandemic teach us it is possible with precautionary way of distancing, being hygienic [95, 96]. Just like the mutations since the early known days of existence of viruses the norms of our living is parallels adopting to the new norms that teach us to overcome and/or adopt to live with it. As the norm keeps to move towards an organized purpose of lives, the unknown and tiny virus force us to adopt and live with it, minimizing the impact of disease transmission. Table 5 shows the new normal that becomes a normal since the span of four industrial revolution [5].

Table 4 Business trends and nature towards the COVID-19 scenario

Industry	Sector	Prime business	COVID 19 business	References
Integral coach factory	Transport	Train bodies	Make shift hospital beds/quarantine center	[70]
SAIC-GM	Vehicles	Vehicle manufacturing	Facemasks, Disinfectants	[80]
DDB ltd	Furniture	Office furniture's	Hygiene hook	[84]
Anheuser Busch	Beverages	Energy drinks/beverages/water	Hand sanitizers, masks	[86]
Diageo India	Liquor		Alcohol-based hand sanitizers	
Pernod/Brewdog	Brewing	Alcoholic beverages	Hand sanitizers	[87]
Shiseido	Personal care	Hair and skin care	Hand sanitizer/Hydro alcoholic gel	[72]
L'oreal	Cosmetics	Hair and skin care	Alcohol-based hand sanitizers/hygiene products	[88]
Ast sportswear	Apparel and fashion	Source clothing, outerwear, leather jackets	Reusable cotton face masks	[89]
Hedley and Bennett		Uniforms and aprons	Washable face masks	
H&M	Fashion	Clothing-retail company	PPE	[90]
Kering	Fashion	Luxury goods	Surgical masks	[91]
PRADA	Fashion	Leather goods/perfumes/fashion accessories	Fashionable and medical face masks as well as hospital gowns	[92]
UNIQLO	Retail	Fabric	Face masks	[93]
UiPath	Automation	Robotics	Software "robots"	[84]
Foxconn	Electronic gadgets	Assembling apple's iPhone	Facemasks	[84]
Apple	Software	Software/Hardware	Foldable plastic face shields	[94]
Slightly robot		Wristband	Smart bands	[84]
Universities globally	Education	Education	Face masks/face shields/ventilators	[73]

Table 5 New norms of the various industrial revolution phases

Industrial Revolution	Major Outbreak	Period	New Normal
	SMALL POX CHOLERA	15TH TO 17TH CENTURY	Vaccination
	1918 SPANISH FLU DENGUE	18TH CENTURY	Sanitization
	HIV AIDS EBOLA NIPAH SARS	19TH -20TH CENTURY	Protections
	ZIKA/YELLOW FEVER SWINE/AVIAN FLU MERS COV (SAUDI) MERS COV (KOREA) EBOLA SARS-COV-2	20TH CENTURY - CURRENT	Social Distancing

With the similarity to that of the coronaviruses such as the SARS, MERS it is expected to develop considerable immunity. Both the infected and asymptotically infected left unnoticed would have developed immunity to such stains of viruses for the future [44].

A second wave or subsequent health challenges are there with the current pandemic to continue for a while. With the economy and livelihood is at the crossroads most of the countries will revert to the normalized mood but with the virus still prevailing. Pharmaceutical industries might need to step in with the modern tools as available to develop solutions towards a newer version as we found COVID-19 stain is evolving when moving from host to host. A formal disease surveillance system is essential to monitor the outbreak of any epidemic in the world [44]. It is important to have one disease reporting system to fetch data from reliable sources like CDC (www.cdcgov.in) [16], WHO (www.who.int) [39], etc.

The advent of technology and open sharing of new information through International collaborations help to limit the spread of the future outbreak in parallel to the newer vaccines into the market [47]. The larger lesson from the scenario is leading us to think out of the box solutions towards education, business, governance and society the COVID-19 has brought us.

Table 6 summarize the new normalcy through the key **SIX** areas towards the operation of a society, at least for the near future.

Table 6 The new normalcy towards the operation of a society post COVID-19

Priority		New normal		Preparedness
Hygiene	Entrance/exit/key locations, escalator handles/lifts	Sanitization and sterilization	Malls, medium to large events, entertainment outlet	• Crowd Management Procedures Mandatory PPE for the frontend sales person including cashiers • SOP for sterilization and sanitization • SOP to handle medical susceptible cases • Restriction on the volume of lift usages • Scenario plan for workers with infection
		Provisions of masks and hand sanitizers at key locations		
		Provisions of hand gloves at grocery shops		
	Rest rooms	Dedicated cleaning staff for—regularly sanitize		
	Retail shops	No return policy at retails		
		Information hoardings/notice on importance of hygiene		
	Throughout	Smart payments or card base payment throughout the visit with limited cash transfers		
	Warehouse	Hygienic supply chain handling and distribution		
Health safety	Entrance and exit	Point of thermal scanning/ CCTV integrated thermal/ Use of drone for open gathering		
	Medic	Isolation and Quarantine Room (Isolated from business area)		
	Medic	In-house medical officer (larger events/malls/outlets) or provision to contact with standby transport service		
	Occupancy	Ceiling slowly increased with time and time limit to occupancy for every visit		

(continued)

Table 6 (continued)

Priority		New normal		Preparedness
Social distancing	Counters	Distancing marks at the cash counter/exchange and ticket counters		
	Elevators/lift/escalators	Elevator capacity reduced. Alternate floors for different elevator		
	Food court	Minimum space between tables. Use of bring your own spoons (BYS) or pay to buy (not recyclable)		
	Parking	Cashless credit parking/no valet parking		
	Air conditioner	Monitor for the air quality through sensors		
	Private transportation	Less contact reserve/credit reserve or app reserve where details are traceable	Construction and renovation sites	
	Construction	Worker to stay at the site it selves with appropriate distancing		
	Shift movement	Attendance would be based on contactless through scan/barcode		
	Office	Work from home would be a norm to reduce mobility	Corporates, Startups, University	SOP appropriate to the industry
	Car hire	Discourage passenger private sharing options, possible buying for personal transport	Rental Services, Auto	
Contact tracing	Earmarked places of infections	Mobile phone would be the hotspot as tracing as a global surveillance	Governmental but within the personal protection act as applicable [97]	

(continued)

Table 6 (continued)

Priority		New normal		Preparedness
		Spots that are RED marked should be a place with more stringent measures		
	Personal mobile phone	Bluetooth would play part in verify on the person nearby		
Travel	Local tourism	Domestic travel would see a sharper growth till middle of 2021 [96]	Tours and travel Industry	Aircraft disinfection SOP SOP for embarkation and disembarkation [75]
		Combined promotional travel would be the norm		
	Long flights	Airlines would run with limited capacity (removal of middle seat to keep up the distance)		
Education	Life-long learning	E-learning/blended learning would be the norm	School/university	
	Home schooling	Rise in homeschooling would be seen	School	
	University	University run in endowment model with a lot of co-working spaces	University/colleges	Assessment and grading system aligned to local educational agencies
		Cloud-based attendance monitoring assessment communication assessment		
		Workshop and hands-on would be at uni. rest all e-learning to contain student numbers		
		Work-based learning where learn what you need is on cards		

5   Summary

A number of lessons learned from the information and the pandemics, we have analyzed some of the good measures that would be the "new norm" of the future as directed below:

1. **Social distancing** is becoming a norm and part of our movements in the future, with the biggest challenge lies in the higher population density countries like India, Vietnam, Bangladesh, Philippines and similar countries
2. **Health and hygienic** takes centerfold with sanitizers, face mask would become a common tool in the daily needs of work, business.
3. **E-Commerce** will be the norm and a contactless cash/credit-based system will remain part of life.
4. **Demand for health workers**, front liners and service sector aligned to health and hygiene would be higher in the new normalcy domain
5. **Automated operations** are to replace the manual operations with electronic doors, scan or bar code-based assess to lifts, elevators, entrance, cashless toll collection, open-air theatre, etc.,
6. The way the **Supply chain SOP** is redefined with the quarantine of contact surface and isolation of product including storing, the disinfecting area would be part of it
7. **Take away and pre-order** of meals would take center hold than dine in at least for the near future
8. Public gatherings, social events, **large meetings**, conferences, international meetings might drive more audiences through **electronic mode**. Electronic live events would become part and parcel immediately
9. All entry points are expected to have thermal sensors, drones used as thermal sensors in larger gatherings might be employed and an immediate isolation point or ward is expected to be in the new norm
10. **Overseas education**, mobility programs and the related business would see a downfall making the localized education demand rise
11. **Employment model** would see a shift towards entrepreneurship or making newer SME opportunity to rise, government machine should leverage on this to increase localized production thereby increasing new employment opportunity
12. The volume of **increase of private cars** is inevitable as the people are worried about the impact of travel through public transport, hence the mobility on the auto industry is to happen
13. The eLearning model though in initial doesn't favor from parents, but slowly it is becoming part of the home. **Home schooling** is to prevail and could slowly replace conventional school systems
14. **Co-working space** would be the norm in the post COVID-19 era, an unconventional approach for modern offices
15. **Entertainment** outlets, theme parks would not be able to see the spike as seen before at least to end of a year

16. **Tourism** would take backstage and the players need to work with other industry to stabilize for the short term till the normalcy is in complete phase by mid of next year.

References

1. Small pox and its prevention. Lancet. **195**(5047), 1129–1132 (1920)
2. De los Reyes, A.A., Escaner, J.M.L.: Dengue in the Philippines: model and analysis of parameters affecting transmission. J. Biol. Dyn. **12**(1), 894–912 (2018)
3. Dixon, S.: The impact of HIV and AIDS on Africa's economic development. BMJ **324**(7331), 232–234 (2002)
4. Danziger, R.: The social impact of HIV/AIDS in developing countries. Soc. Sci. Med. **39**(7), 905–917 (1994)
5. Industrial revolution—from industry 1.0 to industry 4.0. Desoutter Industrial (2020)
6. History of Smallpox. CDC. https://www.cdc.gov/smallpox/history/history.html (2016). Accessed 22 Apr 2020
7. Mitchell, J.: Small pox and AmmaS in South Africa. Lancet **200**(5172), 808–812 (1922)
8. The Relationshiop of small pox and alsatrim. Lancet **196**(5075), 1153–1154 (1920)
9. Whipps, H.: How smallpox changed the world. Live Science (2008)
10. Harris, J.B., LaRocque, R.C., Qadri, F., Ryan, E.T., Calderwood, S.B.: Cholera. Lancet **379**(9835), 2466–2476 (2012)
11. Ivers, L.C.: New strategies for cholera control. Lancet Glob. Heal. **4**(11), e771–e772 (2016)
12. WHO: Global epidemics and impact of cholera. https://www.who.int/topics/cholera/impact/en/ (2020). Accessed 21 Apr 2020
13. Spanish flu: the virus that changed the world. History Extra. https://www.historyextra.com/period/20th-century/spanish-flu-the-virus-that-changed-the-world/ (2020). Accessed 19 Apr 2020
14. Karlsson, M., Nilsson, T., Pichler, S.: The impact of the 1918 Spanish flu epidemic on economic performance in Sweden: an investigation into the consequences of an extraordinary mortality shock. J. Health Econ. **36**, 1–19 (2014)
15. Chowell, G., Bettencourt, L.M., Johnson, N., Alonso, W.J., Viboud, C.: The 1918–1919 influenza pandemic in England and Wales: spatial patterns in transmissibility and mortality impact. Proc. R. Soc. B Biol. Sci. **275**(1634), 501–509 (2008)
16. Pandemic resources. CDC. https://www.cdc.gov/flu/pandemic-resources/1918-pandemic-h1n1.html (2020). Accessed 12 Apr 2020
17. Gubler, D.J.: Dengue and dengue hemorrhagic fever. Clin. Microbiol. Rev. **11**(3), 480–496 (1998)
18. Ryu, W.S.: Influenza viruses. In: Molecular virology of human pathogenic viruses, pp. 195–211 (2017)
19. Brown, R.: The virus detective who discovered Ebola in 1976. BBC News (2014)
20. Ramharack, P., Devnarain, N., Shunmugam, L., Soliman, M.E.S.: Navigating research toward the re-emerging Nipah virus—a new piece to the puzzle. Curr. Pharm. Des. **25**(12), 1392–1401 (2019)
21. Arora, A., Dogra, A., Dogra, A., Goyal, B., Maulik Sharma, A.: Nipah virus: an outbreak of deadly paramyxvirus. Biomed. Pharmacol. J. **11**(3), 1177–1185 2018
22. Hayman, D.T.S.: Nipah virus: a virus with multiple pathways of emergence. In: The role of animals in emerging viral diseases, pp. 293–315 (2014)
23. McKibbin, W., Lee, J.W.: Estimating the global economic costs of SARS, Institute of Medicine, US (2004)
24. Wang, L.-F., Shi, Z., Zhang, S., Field, H., Daszak, P., Eaton, B.: Review of bats and SARS. Emerg. Infect. Dis. **12**(12), 1834–1840 (2006)

25. McKibbin, W.J.: The swine flu outbreak and its global economic impact. Brookings (2009)
26. Jang, H., et al.: Highly pathogenic H5N1 influenza virus can enter the central nervous system and induce neuroinflammation and neurodegeneration. Proc. Natl. Acad. Sci. **106**(33), 14063–14068 (2009)
27. Häfner, S.: Birds of ill omen—is H7N9 the harbinger of the next pandemic? Microbes Infect. **15**(6), 429–431 (2013)
28. Liu, Q., Lu, L., Sun, Z., Chen, G.-W., Wen, Y., Jiang, S.: Genomic signature and protein sequence analysis of a novel influenza A (H7N9) virus that causes an outbreak in humans in China. Microbes Infect. **15**(6–7), 432–439 (2013)
29. Kim, K.H., Tandi, T.E., Choi, J.W., Moon, J.M., Kim, M.S.: Middle east respiratory syndrome coronavirus (MERS-CoV) outbreak in South Korea, 2015: epidemiology, characteristics and public health implications. J. Hosp. Infect. **95**(2), 207–213 (2017)
30. Fung, I.C.-H., et al.: Twitter and Middle East respiratory syndrome, South Korea, 2015: a multi-lingual study. Infect. Dis. Heal. **23**(1), 10–16 (2018)
31. Oldstone, M.B.A., Rose Oldstone, M.: Chapter 8—Ebola's curse: impact on the economics of West Africa. In: Oldstone, M.B.A., Rose Oldstone, M. (eds.) Ebola's curse, pp. 79–86 (2017)
32. Qureshi, A.I.: Ebola virus: the origins. In: Ebola virus disease, pp. 23–37 (2016)
33. Qureshi, A.I.: Chapter 4—Ebola virus disease epidemic in light of other epidemics. In: Ebola virus disease, pp. 39–65 (2016)
34. Qureshi, A.I.: Chapter 13—economic and political impact of Ebola virus disease. In: Ebola virus disease, pp. 177–191 (2016)
35. Snyder, B.: Ebola outbreak's impact on 5 key industries. Fortune (2014)
36. Zaki, A.M., van Boheemen, S., Bestebroer, T.M., Osterhaus, A.D.M.E., Fouchier, R.A.M.: Isolation of a novel coronavirus from a man with pneumonia in Saudi Arabia. N. Engl. J. Med. **367**(19), 1814–1820 (2012)
37. Petersen, E., Hui, D.S., Perlman, S., Zumla, A.: Middle east respiratory syndrome—advancing the public health and research agenda on MERS-lessons from the South Korea outbreak. Int. J. Infect. Dis. **36**, 54–55 (2015)
38. Bauch, C.T., Oraby, T.: Assessing the pandemic potential of MERS-CoV. Lancet **382**(9893), 662–664 (2013)
39. WHO the cumulative number of reported probable cases of severe acute respiratory syndrome (SARS), World Health Organization (2020)
40. Symptoms of coronavirus. CDC. https://www.cdc.gov/coronavirus/2019-ncov/symptoms-testing/symptoms.html (2020). Accessed 16 Apr 2020
41. Naughton, J.: When covid-19 has done with us, what will be the new normal? The Guardian (2020)
42. Zhou, P., et al.: A pneumonia outbreak associated with a new coronavirus of probable bat origin. Nature **579**(7798), 270–273 (2020)
43. Poon, L.L.M., et al.: Early diagnosis of SARS coronavirus infection by real time RT-PCR. J. Clin. Virol. **28**(3), 233–238 (2003)
44. Li, J.-Y., et al.: The epidemic of 2019-novel-coronavirus (2019-nCoV) pneumonia and insights for emerging infectious diseases in the future. Microbes Infect. **22**(2), 80–85 (2020)
45. Kofi Ayittey, F., Dzuvor, C., Kormla Ayittey, M., Bennita Chiwero, N., Habib, A.: Updates on Wuhan 2019 novel coronavirus epidemic. J. Med. Virol. **92**(4), 403–407 2020
46. Keogh-Brown, M.R., Smith, R.D.: The economic impact of SARS: how does the reality match the predictions? Health Policy. **88**(1), 110–120 (2008)
47. Pan, X., Gao, T., Li, Z., Pan, C., Pan, C.: Lessons learned from the 2019-nCoV epidemic on prevention of future infectious diseases. Microbes Infect. **22**(2), 86–91 (2020)
48. Kwok, K.O., Tang, A., Wei, V.W.I., Park, W.H., Yeoh, E.K., Riley, S.: Epidemic models of contact tracing: systematic review of transmission studies of severe acute respiratory syndrome and middle east respiratory syndrome. Comput. Struct. Biotechnol. J. **17**, 186–194 (2019)
49. Bonilla-Aldana D.K., et al.: SARS-CoV, MERS-CoV and now the 2019-novel CoV: have we investigated enough about coronaviruses?—a bibliometric analysis. Travel Med. Infect. Dis. (2020)

50. Interim guidelines for collecting, handling, and testing clinical specimens from persons for coronavirus disease 2019 (COVID-19). Centers for Disease Control and Prevention. https://www.cdc.gov/coronavirus/2019-ncov/lab/guidelines-clinical-specimens.html (2020). Accessed 17 Apr 2020
51. Chong, Y.L., Ahmed, S., Tan, W.L.G.: Surgical response to COVID-19 pandemic: a Singapore perspective. J. Am. Coll. Surg. (2020)
52. Thoracic Surgery Outcomes Research Network, I.: COVID-19 guidance for triage of operations for thoracic malignancies: a consensus statement from thoracic surgery outcomes research network. Ann. Throacic Surg. (2020)
53. YauOng, C.: Ongoing lessons during the COVID-19 pandemic. J. Am. Coll. Surg. (2020)
54. Ducournau, P.A.L.F., Arianni, M., Awwad, S., Baur, E.M., Beaulieu, J.Y., Bouloudhnine, M., Caloia, M., Chagar, K., Chen, Z., Chin, A.Y., Chow, E.C., Cobb, T., David, Y., Delgado, P.J., Woon Man Fok, M., French, R., Golubev, I., Haugstvedt, J.R., Ichihara, S., Jorquera, R.A., Koo, S.C.J.J., Lee, J.Y., Y.K.: COVID-19: initial experience of an international group of hand surgeons. Hand Surg. Rehabil. **39**, 159–166 (2020)
55. Downsizing pandemic through the measures for the COVID19 outbreak in Malaysia (2020)
56. Zhong, L., Mu, L., Li, J., Wang, J., Yin, Z., Liu, D.: Early prediction of the 2019 novel coronavirus outbreak in the mainland China based on simple mathematical model. IEEE Access **8**, 51761–51769 (2020)
57. Mims, C.: Chapter 10—the threat of new diseases. War Within Us (2000)
58. Hosein, R., Khadan, J., Paul, N.: An assessment of the factors determining medal outcomes at the Beijing olympics and implications for CARICOM economies **62**(1–2) (2013)
59. Ha, K.M.: Emergency response to the outbreak of COVID-19: the Korean case (2020)
60. Decks cleared for rapid testing in 170 hotspots across country, testing ramped up, 5 lakh kits arrive from China, another 6. 5 lakh on way. Indian Express (1932)
61. WTO: Impact assessment of the COVID-19 outbreak on international tourism (2020)
62. KPMG: Potential impact on COVID-19 on the Indian economy, India (2020)
63. Simon Burgess, H. H. S.: Schools, skills, and learning: the impact of COVID-19 on education. Vox (2020)
64. Opinion: COVID-19 impact on Indian automotive industry—taking action in troubled times. ET Auto (2020)
65. Becker, D.: COVID-19 impact on the automotive sector. KPMG Blog. https://home.kpmg/xx/en/blogs/home/posts/2020/03/covid-19-impact-on-the-automotive-sector.html. Accessed 22 Apr 2020
66. "Covid-19 Is Bad for the Auto Industry—and Even Worse for EVs," *Wired*, 2020. [Online]. Available: https://www.wired.com/story/covid-19-bad-for-auto-industry-worse-for-evs/ (2020). Accessed 22 Apr 2020
67. ICIRAST: Containing COVID19 impacts on Indian agriculture (2020)
68. Dev, M.: Addressing COVID-19 impacts on agriculture, food security, and livelihoods in India, International Food Policy Research Institute (2020)
69. Singh, I. S.: Agriculture in the time of Covid-19. Agri Business (2020)
70. Sivapriyan, E.: Coronavirus: ICF to open up its factory to produce ventilators. Deccan Herald (2020)
71. Ford to produce respirators, masks for COVID-19 protection in Michigan; scaling up production of gowns, testing collection kits. Ford. https://corporate.ford.com/articles/products/ford-producing-respirators-and-masks-for-covid-19-protection.html (2020)
72. Shiseido develops hand sanitizer to curb COVID-19 spread. GCI Magazine (2020)
73. Taylor's Me.reka makerspace seeks partners to increase face shield production to equip frontliners. Taylor's Uni. https://university.taylors.edu.my/en/campus-life/news-and-events/news/taylors-mereka-makerspace-seeks-partners-increase-face-shield.html (2020). Accessed 24 Mar 2020
74. Coronavirus: iPhone manufacturer Foxconn to make masks. BBC News (2020)
75. Covid-19: AirAsia implements new travel policies. The Star. https://www.thestar.com.my/business/business-news/2020/04/27/covid-19-airasia-implements-new-travel-policies (2020). Accessed 28 Apr 2020

76. Taylor, C.: Mater hospital turns to robots to help in battle against Covid-19. The Irish Times
77. Ilancheran, M.: COVID-19, medical drones, & the last mile of the pharma supply Chain. Pharmaceutical Online (2020)
78. Covid-19 pandemic: Uber ties up with Flipkart to deliver essential supplies to customers. Economic Times (2020)
79. New balance to manufacture face masks for healthcare workers amid coronavirus. Bleacher report (2020)
80. Monica: Automaker SAIC-GM-Wuling rolls out first self-made face masks. Auto News (2020)
81. Johnson and Johnson: 5 impactful ways Johnson & Johnson is helping in the fight against covid-19. https://www.jnj.com/latest-news/ways-johnson-johnson-helping-to-fight-covid-19 (2020). Accessed 23 Apr 2020
82. GM mobilizes to support communities amid coronavirus pandemic. General Motors. https://media.gm.com/media/us/en/gm/news.detail.html/content/Pages/news/us/en/2020/apr/0409-coronavirus-update-11-support.html (2020) Accessed 22 Apr 2020
83. Volswagenag: From making cars to ventilators. https://www.volkswagenag.com/en/news/2020/03/cars_to_ventilators.html (2020). Accessed 22 Apr 2020
84. Josephine Mason, L.I., Henderson, P.: Mother of invention: the new gadgets dreamt up to fight coronavirus. Thomson Reuters (2020)
85. Virgin: Virgin orbit develop and design mass producible ventilators for COVID-19 patients. https://www.virgin.com/news/virgin-orbit-develop-and-design-mass-producible-ventilators-covid-19-patients (2020). Accessed 24 Apr 2020
86. Covid-19 impact: Budweiser maker equips 15 lakh frontline workers with masks and hand sanitizers. ET Retail (2020)
87. Jacobsen, J.: Spirits community rallies to help address hand sanitizer shortage. Beverage Industry (2020)
88. Loreal: COVID-19: our global solidarity efforts. https://www.loreal.com/group/our-activities/covid-19-our-global-solidarity-efforts (2020). Accessed 24 Apr 2020
89. Schmidt, I.: Fashion brands are making face masks, medical gowns for the coronavirus crisis. Los Angeles Times (2020)
90. Glover, S.: Brands inc Primark and H&M back COVID-19 plan. Ecotextile News (2020)
91. Kering: Kering contributes to the fight against COVID-19. https://www.kering.com/en/news/kering-contributes-to-the-fight-against-covid-19 (2020)
92. Innovation in the time of covid-19. The Straits Times (2020)
93. Coronavirus (COVID-19): impact on fashion and luxury in China. IFA Paris (2020)
94. Coronavirus: Apple and Tesla reveal the new products they're making in COVID-19 fight. ZDNet (2020)
95. The coronavirus will hit the tourism and travel sector hard. The Conversation (2020)
96. Flynn, D.: Post-coronavirus, 'normal' travel may not resume until 2023. Executive Traveller. https://www.executivetraveller.com/news/post-coronavirus-normal-travel-may-not-resume-until-2023 (2020). Accessed 18 Apr 2020
97. Darrell Etherington: MIT develops privacy-preserving COVID-19 contact tracing inspired by Apple's 'Find My' feature. Techcrunch. https://techcrunch.com/2020/04/09/mit-develops-privacy-preserving-covid-19-contact-tracing-inspired-by-apples-find-my-feature/ (2020). Accessed 12 Apr 2020

A Hybrid Method of MCDM for Evaluating Financial Performance of Vietnamese Commercial Banks Under COVID-19 Impacts

Phi-Hung Nguyen, Jung-Fa Tsai, Yi-Chung Hu, and G. Venkata Ajay Kumar

Abstract As an essential and exciting topic in financial management, MCDM has been widely used in evaluating financial performance to improve the suitability and reliability of financial indicators with respect to the impacts of both qualitative and quantitative information. This chapter aims to present a hybrid MCDM approach to evaluate the Vietnamese banking sector's performance under COVID-19 impacts. The proposed method utilizes The Criteria Importance Through Inter-criteria Correlation (CRITIC) technique to determine objective weights of financial ratios. Decision-Making Trial and Evaluation Laboratory (DEMATEL) method is employed to obtain the cause-effect relationship and the subjective weights based on experts' judgments. Bank alternatives' ranking is estimated using the Technique for Order Preference by Similarity to Ideal Solution (TOPSIS) approach. The empirical data of 23 Vietnamese commercial banks gathered from 2019-Q3/2020 is illustrated. The results of this chapter show that the COVID-19 pandemic has significant effects on the quality of assets, the liquidity of banks, and the growth rate. When the economy slows down, banks face the challenge of keeping up with demand and raising additional resources to balance the situation. For better intelligent risk management systems, faster digital transformation is needed for the Vietnamese banking system. Furthermore, the proposed method offers useful insights and is

P.-H. Nguyen (✉)
Department of Business Management, National Taipei University of Technology, Taipei 10608, Taiwan
e-mail: hungnp30@fe.edu.vn

Faculty of Business, FPT University, Hanoi 100000, Vietnam

J.-F. Tsai
Department of Business Management, National Taipei University of Technology, Taipei 10608, Taiwan
e-mail: jftsai@ntut.edu.tw

Y.-C. Hu
Department of Business Administration, Chung Yuan Christian University, Taoyuan City 32023, Taiwan

G. V. Ajay Kumar
Kadapa, India

© Institute of Technology PETRONAS Sdn Bhd 2022
S. A. Abdul Karim (eds.), *Shifting Economic, Financial and Banking Paradigm*,
Studies in Systems, Decision and Control 382,
https://doi.org/10.1007/978-3-030-79610-5_2

applicable to many businesses, proving helpful in dealing with complex criteria problems for stakeholders.

Keywords Performance evaluation · Banking industry · COVID-19 · MCDM · DEMATEL · TOPSIS · Vietnam

1 Introduction

The COVID-19 pandemic outbroke at the beginning of 2020, causing devastating effects on the world's economy. In reaction to changes in social and economic environments, governments proposed a series of supporting provisions to minimize such adverse effects. The primary purpose is to sustain jobs, support non-financial entities and preserve spearhead business sectors. To extinguish the negative spatial impacts in the commercial paper market, equity market, stock market, the US central bank has taken instant measures consisting of three classifications: monetary policy, liquidity procedure, and credit programs. While monetary and liquidity provisions aim to set the policy rates and penalty rates to lower bounds, along with extensive asset purchase programs, the Federal Reserve, backed by a $450 billion worth first-loss protection scheme, helps maintain market liquidity [1].

The global dissemination of coronavirus (COVID-19) has been dramatically affecting financial markets around the world. It has generated an extraordinary risk of risk, resulting in substantial losses for investors in the short-run [2, 3]. Most Vietnamese financial institutions were affected by the COVID-19 pandemic in the short-run, long-run [4].

In the Vietnamese banking sector, managers and academics pay considerable attention to enhance service quality and financial performance because of the significant contribution to the annual GDP [5]. The financial analysts expect that virtually the banks' deposit rates, interest rates, and debt repayments would fall over time [6]. The demand for short-term loans from households has declined substantially, while the demand for non-financial companies has improved. The pandemic has led to isolated individuals, slower usage, and consideration in credit card loans. Since microenterprise owners return to work to avoid high-interest rates for early repayment, corporations' loans are diminished in their ability to restart. During this time, the services sector and small and medium-sized enterprises (SMEs) struggled to survive. The pandemic has had various impacts on the quality of bank reserves. Vietnamese manufacturing, marketing, importing, and exporting activities have been affected by the pandemic. SMEs are perceived as riskier, and small business finances are viewed as less secure in terms of company sizes and credit risks [7].

The Vietnamese State bank, through reducing operating interest rates, instructing commercial banks to restructure loans, lower interest rates, and postpone debts for customers facing difficulties during the pandemic, has achieved temporary successes in containing the virus's harmful impacts. Consequently, Vietnam's GDP continued to grow at 3.82% in the first and 0.36% in the second quarter of 2020. There was a small negative growth rate among several countries during the pandemic, which means this is a positive sign. The banks, therefore, play an essential yet indispensable role when the economy stagnates.

Regarding the importance of the banking system, analyzing financial ratios' performance sets entails substantial academic attention. Investigating the financial performance of any organization has gained interest from a diverse group of participants, including managers, creditors, financial experts, prospective investors, researchers [8–12].

A shred of increasing evidence worldwide suggests that well-developed and comprehensive banking systems are essential for rapid growth. The banking sector is incorporated into the country's central money market. Evaluation of banking efficiency is critical for growth and profitability in a dynamic market structure. Productivity tests an organization's success and is often necessary to observe progress. Banks are required to take the lead in choosing appropriate actions. However, taking clear performance metrics into account when selecting the best bank can be helpful for a variety of reasons:

Firstly, performance evaluation can help the bank manager assign more appropriate tasks and formulating more effective plans and policies. Bank executives are concerned about rising productivity to enhance their benefits and service quality.

Secondly, Multi-criteria decision-making (MCDM) methods are widely employed to overcome the shortcomings of traditional analyzing methods because they simultaneously consider multiple dependent and independent attributes. Furthermore, the MCDM model is utilized to measure, rate, and benchmark efficiency while also searching for the optimal solution.

Owing to the existence of multiple input and output and the massive impacts of the COVID- 19, the MCDM techniques could be applied to calculate performance and benchmarking in the Vietnamese banking sector. The main research question of these chapters is described as follows:

- Which criteria is proposed to evaluate financial performance in the Vietnamese banking industry under COVID-19 impacts?
- How could the importance of the selected criteria (weights) be found by an integrating CRITIC and DEMATEL method?
- How could a hybrid MCDM method measure and rank the financial performance in the Vietnamese banking industry under COVID-19 impacts?

This chapter is organized into five sections. Section 2 presents the comprehensive existing literature on banks' performance evaluation. In the next section, the proposed financial criteria are briefly explained. The fourth section summarizes the procedure of proposed MCDM methods. Section 4 illustrates an empirical case with real data. Finally, conclusions are drawn.

2 Literature Review

Financial analysis is the scientific process of evaluating and measuring relationships between financial attributes on financial statements to make decisions [13]. In today's world, the banking sector and banks play a crucial role in the industry and economy of countries [14]. As a financial institution that manages businesses and investments, the bank's value assessed includes various segments of society. In the field of financial analysis, applications of CAMELS model, Dupont system, and Discounted Cash Flow method are widespread [15–24].

As studied by [25], standard success rankings are based on financial data that is clear and reliable, such as profitability ratios, liquidity ratios. Performance assessment should be consistent with the company's strategic approach and have a set of comprehensive metrics (both financial and non-financial indicators) that an entity can use to produce the optimal outcomes in its projects, investments, and acquisitions.

Benjamin et al. empirically studied dividend policy using a Dupont-based analysis in Malaysia [17]. On the other hand, as shown in [21], financial leverage was considered the most relevant indicator for analyzing ROE ratios. References [24] used the discounted cash flow technique to assess hypothetical monoculture and IMTA operations' profitability over ten years in Canada. The balanced scorecard analysis should be incorporated into performance models to recognize effective performance indicators and those that will help design and execute management practices [26–28].

Given the ongoing debate about whether conventional financial ratios are still the best way to measure a company's results, The CAMELS model is also commonly considered to assess bank efficiency. The differences of opinion between the priorities based on the probability of bankruptcy and bank results and the differences of view based on the demographic characteristics of the experts are also examined. Supervisory authorities have adopted the CAMELS model in many countries. In a recent study, Pekkaya and Demir identified the CAMELS model's most significant bank performance dimensions. Their findings indicated that Asset is the most significant dimension, followed by Earnings, Liquidity, and Management [20]. The researchers, Rashid and Jabeen, reviewed the factors affecting Islamic and commercial banks in Pakistan; and they developed the CAMELS ratios to quantify them [18].

The literature investigation also found that researchers generally use ratio analysis and Data Envelopment Analysis (DEA) to calculate efficiency and compare alternatives in various research fields [29]. Different models have been used in evaluating output, and more complicated business operations have spurred on DEA extensions and improvements [30, 31]. Regarding the research of [32], they investigated 80 published papers, which were applied the DEA model about the banking sector in 24 countries. Wang et al. used the DEA method to examine, assess, and test 16 major Chinese commercial banks' efficiencies in 2003–2011 [33]. In other research of [34], they used the network-DEA centralized efficiency model. Their results showed that Brazilian banks differ based on cost efficiency and others on efficient efficiency. Meanwhile, DEA has two models: the Charnes-Cooper-Rhodes model with a persistent return to scale assumption and Banker-Charnes-Cooper model with a variable return to scale [35]. While standard DEA models and variations have been commonly used to research banking, these models do not account for the internal structure in terms of measures that characterize banking operations efficiency.

The financial evaluation process is more complex and mostly inconsistent, not to mention the objectivity and subjectivity of decision-makers. Under this condition, financial analysis is conducted in diverse areas of study. MCDM methods play a role in providing decision guidance for decision-makers [36–40]. As described above, each MCDM approach has its point of view in assessing and ranking alternatives. Dinçer and Yüksel evaluated new service development competencies in the Turkish banking sector [41]. Zhao et al. researched the challenges brought by Fintech startups [42]. As studied by [43], an integrated BSC and fuzzy MCDM model for banking performance evaluation was presented.

However, the aforementioned models' structures are discrete, limiting broadband understandings of financial analysis. Meanwhile, judging by the MCDM approach's extensive grey programming methods in solving various issues, the method and its upgraded versions have been quite popular in diverse research areas. References [14, 41, 42, 44–49]. Additionally, very few studies on financial analysis with MCDM applications have been conducted. For this reason, adopting an MCDM based method to financially assess the COVID-19 performance of the Vietnamese banking industry is dependable.

DEMATEL has been used in several experiments over the last decade, and several different versions have been put forward in the literature. DEMATEL is an efficient tool for defining the cause-effect components of a complex system [50–55]. It deals with evaluating interdependent relationships among factors and finding the critical ones through a visual structural model. Risk management helps improve company performance, which leads to the ability to make more confident supply chain sustainability decisions. Because of the size of the enterprise, the risk management techniques of the sustainable supply chain must be more advanced for large companies. In the study of [56, 57], a combination of TOPSIS and CRITIC is proposed to investigate risk management problems. In [58], objective weights resulted from the CRITIC method to assess Third-Party Logistics.

3 A Hybrid MCDM Model for Vietnamese Banking Sector Under COVID-19 Impact

Although both CRITIC and DEMATEL were widely used in literature, there is a lack of research integrating them to determine the objective and subjective weights.

In contrast to the above approaches, this chapter pioneers proposing a hybrid MCDM model of CRITIC–DEMATEL- TOPSIS to explore the Vietnamese banking sector's performance from 2019 to Q1/2020 COVID-19 impacts. The research framework is presented in Fig. 1.

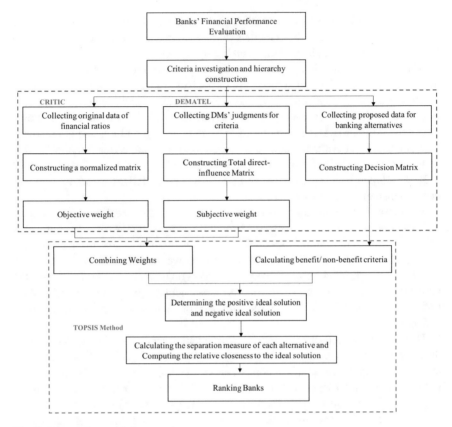

Fig. 1 Proposed research framework

3.1 Objective Weight Method of CRITIC

The Criteria Importance Through Intercriteria Correlation (CRITIC) method was firstly used by Diakoulaki [59]. This objective weighting method is combined by the standard deviation (σj) and correlation coefficient.

Step 1: Normalizing decision matrix using Eqs. (1) and (2):

$$x_{ij} = \frac{r_{ij} - r_i^-}{r_i^+ - r_i^-}; \ i = 1, \ldots, m, \ j = 1, \ldots, n \tag{1}$$

$$x_{ij} = \frac{r_{ij} - r_i^+}{r_i^- - r_i^+}; \ i = 1, \ldots, m, \ j = 1, \ldots, n \tag{2}$$

Step 2: Calculating standard deviation values of r_j (Eq. 3)

$$\sigma_j = \sqrt{\frac{1}{n-1} \sum_{j=1}^{n} \left(x_{ij} - \bar{x}_j\right)^2} \, i = 1, \ldots, m \tag{3}$$

Step 3: Calculating the correlation coefficient between the vectors r_j and r_k (Eq. 4).

$$\rho_{jk} = \sum_{i=1}^{m} \left(x_{ij} - \bar{x}\right)(x_{ik} - \bar{x}_k) / \sqrt{\sum_{i=1}^{m} \left(x_{ij} - \bar{x}_j\right)^2 \sum_{i=1}^{m} (x_{ik} - \bar{x}_k)^2} \tag{4}$$

$$c_j = \sigma_j \sum_{k=1}^{n} \left(1 - \rho_{jk}\right); \; j = 1, \ldots, n \tag{5}$$

Step 4: Equation (5) represents a measure of the conflict created by criterion j through the multiplicative aggregation in Eq. 5:

Step 5: Normalized objective weights are derived by (Eq. 6).

$$w_j = \frac{c_j}{\sum_{j=1}^{n} c_j}; \; j = 1, \ldots, n \tag{6}$$

3.2 Subjective Weight Method of DEMATEL

DEMATEL technique was developed to obtain the relationship of cause-effect criteria and to reflect casual relationships diagram [60].

Step 1: K direct-relation matrix from each expert's judgment. The average direct relation matrix is taken by dividing k experts in Eq. (7):

$$Z = \sum_{i=1}^{k} Z^k / k \tag{7}$$

Step 2: Normalizing matrix A by Eqs. (8) and (9).

$$k = 1 / \max \sum_{j=1}^{n} a^{ij} \tag{8}$$

$$X = K \times A \tag{9}$$

Step 3: Sum of rows (D_i) and the sum of columns (R_i) are calculated by Eqs. (10) and (11):

$$D = \left| \sum_{j=1}^{n} m_{ij} \right|_{n \times 1} \tag{10}$$

$$R = \left| \sum_{i=1}^{n} m_{ij} \right|_{1 \times n} \tag{11}$$

Step 4: Creating the causal relationship diagram based on $(R_i + D_i)$, $(R_i - D_i)$.

Step 5: Weighting criteria based on the value of $(R_i + D_i)$, $(R_i - D_i)$ by Eqs. (12) and (13):

$$W_i = \sqrt{\left[(R_i + D_i)^2 + (R_i - D_i)^2\right]} \tag{12}$$

$$W_i^{nor} = \frac{w_i}{\sum_{i=1}^{n} w_i}; \text{ where } W_i^{nor} \text{ is normalized weights of criteria} \tag{13}$$

3.3   Ranking Method of TOPSIS

TOPSIS method was presented to compare alternatives by analyzing the similarity with the ideal point. The optimum value is closer to the flattering- ideal and further away from the reference solution. TOPSIS method is presented as follows:

Step1: Normalizing decision matrix by Eq. (13).

$$R_{ij} = \frac{m_{ij}}{\sqrt{\sum_{j=1}^{M} m_{ij}^2}} \tag{13}$$

Step 2: Multiplying the weights w_i with the normalized decision matrix r_{ij} in Eq. (14).

$$V_{ij} = w_j \, R_{ij} \tag{14}$$

Step 3: Defining the positive ideal solution and negative ideal solution using Eq. (15).

$$
\begin{aligned}
A^+ &= \{V_{ij}^+ 1, \ldots\ldots V_n^+\} = \{(\max V_{ij} \mid i \in I'), (\min V_{ij} \mid i \in I'')' \\
A^- &= \{V_{ij}^- 1, \ldots\ldots V_n^-\} = \{(\max V_{ij} \mid i \in I'), (\min V_{ij} \mid i \in I'')'
\end{aligned} \tag{15}
$$

Step 4: Computing separation measure of each alternative with Eqs. (16) and (17).

$$D_j^+ = \sqrt{\sum_{i=1}^{n} \left(v_{ij} - v_j^+\right)^2} \tag{16}$$

$$D_j^- = \sqrt{\sum_{i=1}^{n} \left(v_{ij} - v_j^-\right)^2} \tag{17}$$

Step 5: Computing relative closeness to the ideal solution (CCi) in Eq. (18).

$$C_{ci} = \frac{D_j^-}{D_j^+ + D_j^-} \tag{18}$$

Step 6: Ranking the preference order according to the descending order of (C$_{Ci}$).

4 An Empirical Analysis

MCDM approach has long been a popular path of study in diverse sectors; as an effective technique to help identify causality over-complicated criteria, it has gained more attention in recent years. However, applying the MCDM and its variations in financially assessing the Vietnamese banking sector is unprecedented. This study seeks to fill the research gap and proposed an integrated MCDM methodology to show the interrelationship of bank financial results during the COVID-19 pandemic. For the survey, financial experts, scholars were advised to conduct the matrix filling procedure. It should be noted that while financial experts are keen observers of the financial market, they, therefore, have profound knowledge in finance; scholars, on the other, are those having interested in this area of study.

4.1 Proposed Financial Performance Ratios

This chapter is constructed with two different kinds of datasets: financial performance ratios and banking alternatives. The data of the 23 commercial banks listed in the Vietnamese Stock Exchange Market is obtained from the databases of Vietdata (https://finance.vietdata.vn/). For the analysis, a set of variables extracted from the financial statements. Five dimensions (Asset Quality, Growth Rate, Liquidity Ratios, Profitability, Valuation ratios) and 24 proposed financial ratios have been selected with the defined keywords based on literature review from the ScienceDirect (https://www.sciencedirect.com/). Table 1 illustrates brief explanations, proposed criteria for the Vietnamese Banking Industry.

4.2 Weighting Results of CRITIC and DEMATEL

Regarding DEMATEL results, it can be seen that negative (R–C) values consist of G1, G2, G8, G10, LR3, LR4, LR5, whereas the remaining factors are cause ratios. Among these relationships, G1, G2, LR4, LR5, PR1, PR2, PR3 are significant influencing ratios (Table 2). This is evident by the causal relationship diagram of criteria,

Table 1 Proposed financial ratios for vietnamese banking industry

Dimension	Criteria	Definitions	Keywords	Literature
Asset quality	AQ1	Bad debt provision/ Total bad debt	Provision to total bad debt	
	AQ2	Loan under follow-up (gross)/ Total Loans	Non-performing Loans (NPL)	[67, 68]
Growth rate	G1	Growth of total assets	Total assets	[69, 70]
	G2	Growth of owner's equity	Owner's equity	[71, 72]
	G3	Growth of before provision outstanding loans	Outstanding loans	[73, 74]
	G4	Growth of other before other credit institutions outstanding loans	Credit institutions outstanding loans	Proposed by authors
	G5	Growth of deposits from customers	Deposits from customers	[75, 76]
	G6	Growth of deposits and loans from other credit institutions (Interbank mobilization)	Deposits and loans from other credit institutions	[77, 78]
	G7	Growth of net interest income	Net interest income	[79, 80]
	G8	Growth of Net fee and commission income	Net fee and commission income	[81, 82]
	G9	Growth of profit before tax	Profit before tax	[83, 84]
	G10	Growth of profit after tax	Profit after tax	[85, 86]
Liquidity ratios	LR1	Outstanding loans (customers)/ Total Assets	Customer loans, total assets	[87, 88]
	LR2	Outstanding loans (other institutions)/ Total Assets	Outstanding loans, total assets	Proposed by authors
	LR3	Security & Investing activities/ Total assets	Security activities, Investing activities, Total Assets	Proposed by authors
	LR4	Outstanding loans/ Customer deposits	Loan to Deposit Ratio (LDR)	Proposed by authors
	LR5	Equity/ Total assets	Equity Ratio	[89, 90]

(continued)

Table 1 (continued)

Dimension	Criteria	Definitions	Keywords	Literature
Profita-bility	PR1	(Investment return – interest paid)/Average assets	Net interest margin (NIM)	[91, 92]
	PR2	Net income / Total assets	Return on assets (ROA)	[93, 94]
	PR3	Net income/ Average common stock equity	Return on equity (ROE)	[95, 96]
	Pr4	Net interest income / Operating income	Net interest income to operating income ratios	Proposed by authors
	Pr5	Non-interest income from service operating/ Operating income	Non-interest income from service working to operating income ratios	Proposed by authors
Valuation ratios	VA1	Net income/Average outstanding shares of the company	Earning per share (EPS)	[96, 97]
	VA2	(Shareholder's equity – Preferred equity)/ Average outstanding shares of the company	Book value per share (BVPS)	[98, 99]

demonstrating a multi-direction arrow for the mentioned cause ratios. Regarding the weight rankings, it is indicated that Vietnamese banks pay more concern to the Asset quality aspect in time of a pandemic. Judging by the high ranks for Asset Quality criteria, the inference here is that these ratios should be prioritized for enhancement. The remaining significant ratios fall in Growth Ratios, Liquidity Ratios, and Profitability aspects, meaning that sustaining growth, maintaining high liquidity, and stabilizing profitability should be done while managing asset quality. The subjective weights (W_{Sub}) of the proposed criteria are presented in Table 2, perspectively.

Objective weights (W_{Obj}) are calculated from the actual financial data of each criterion for each quarterly data (Table 3).

Based on the objective weights W_{Obj} and subjective weights W_{Sub}, the combining weights W_{Comb} are calculated for each quarter (Table 4).

4.3 Ranking Results of TOPSIS

The findings of this chapter show that the effect of the CO19 shock on banks was much more substantial and long-lasting than on non-financial corporations.

According to Table 5, the TOPSIS rankings indicated that Nam A, OCB, TPBank, STB, VIB, MBB, VCB, TCB, Bac A Bank, and LIEN VIET outperformed the

Table 2 Cause/effect relationship and subjective weights

Criteria	R	C	R + C	R–C	W_Sub	Rank
AQ1	9.19	4.90	14.10	4.29	0.04330	2
AQ2	8.89	5.75	14.64	3.14	0.04400	1
G1	6.79	7.49	14.28	**−0.70**	0.04199	5
G2	6.39	7.69	14.08	**−1.30**	0.04154	12
G3	7.34	6.81	14.16	0.53	0.04162	8
G4	7.17	6.89	14.06	0.27	0.04132	16
G5	6.99	6.96	13.95	0.03	0.04099	23
G6	7.04	6.96	14.00	0.09	0.04114	21
G7	7.27	7.20	14.47	0.07	0.04251	3
G8	6.74	7.27	14.01	**−0.52**	0.04120	19
G9	7.10	7.05	14.15	0.05	0.04159	10
G10	6.95	7.19	14.14	**−0.24**	0.04157	11
LR1	7.18	6.97	14.15	0.21	0.04159	9
LR2	7.11	6.94	14.06	0.17	0.04131	18
LR3	6.89	6.94	13.83	**−0.06**	0.04064	24
LR4	6.99	7.25	14.24	**−0.27**	0.04186	7
LR5	6.39	7.53	13.93	**−1.14**	0.04106	22
PR1	6.90	7.43	14.33	**−0.53**	0.04212	4
PR2	6.24	7.70	13.94	**−1.46**	0.04119	20
PR3	6.17	7.98	14.15	**−1.81**	0.04193	6
PR4	7.06	7.02	14.08	0.04	0.04137	15
PR5	6.82	7.24	14.06	**−0.42**	0.04132	17
VA1	7.05	7.05	14.10	0.00	0.04144	13
VA2	6.81	7.27	14.08	**−0.46**	0.04141	14

remainder, dominating the rankings in the first quarter of 2019. However, the orders experienced dramatic changes in the coming periods as the top-performing banks, namely Nam A, OCB deteriorated significantly to make way for LIEN VIET, VIB in the second quarter, while the others in the top 10 were as follows: SGB, VCB, VPB, ACB, ABB, TPBank, TCB, and MBB. This trend was then continued in Q3 and Q4, 2019, where the leading positions were changed much.

Besides realizing the COVID-19 pandemic's devastating effects, the TOPSIS technique was employed using a new integration of objective and subjective weighting techniques to prioritize Vietnamese banks' performances in 2020. More specifically, a data set related to the quarterly operations of selected banks was incorporated to calculate the combining CRITIC – DEMATEL weighting approach based on experts' judgments. This process enables us to comprehensively and rationally appraise the banks' operations during COVID - 19 pandemic.

Table 3 Objective weight results

Objective W	AQ1	AQ2	G1	G2	G3	G4	G5	G6	G7	G8	G9	G10
WQ1/19	0.0545	0.0480	0.0380	0.0392	0.0345	0.0391	0.0357	0.0404	0.0272	0.0370	0.0406	0.0392
WQ2/19	0.0522	0.0441	0.0389	0.0428	0.0359	0.0416	0.0368	0.0369	0.0276	0.0368	0.0473	0.0421
WQ3/19	0.0602	0.0391	0.0388	0.0432	0.0353	0.0421	0.0361	0.0343	0.0349	0.0418	0.0456	0.0439
WQ4/19	0.0509	0.0406	0.0408	0.0424	0.0348	0.0435	0.0345	0.0394	0.0332	0.0391	0.0439	0.0436
WQ1/20	0.0522	0.0423	0.0378	0.0433	0.0351	0.0415	0.0345	0.0340	0.0401	0.0426	0.0399	0.0385
WQ2/20	0.0533	0.0548	0.0492	0.0403	0.0366	0.0410	0.0335	0.0342	0.0325	0.0418	0.0449	0.0458
WQ3/20	0.0494	0.0449	0.0393	0.0379	0.0346	0.0412	0.0351	0.0333	0.0314	0.0381	0.0399	0.0358

Objective W	LR1	LR2	LR3	LR4	LR5	PR1	PR2	PR3	PR4	PR5	VR1	VR2
WQ1/19	0.0543	0.0597	0.0429	0.0430	0.0508	0.0378	0.0451	0.0308	0.0379	0.0410	0.0381	0.0450
WQ2/19	0.0563	0.0582	0.0406	0.0416	0.0505	0.0333	0.0449	0.0286	0.0383	0.0422	0.0378	0.0446
WQ3/19	0.0532	0.0601	0.0402	0.0393	0.0442	0.0358	0.0387	0.0273	0.0382	0.0430	0.0390	0.0458
WQ4/19	0.0534	0.0628	0.0374	0.0409	0.0466	0.0368	0.0397	0.0296	0.0387	0.0447	0.0389	0.0437
WQ1/20	0.0570	0.0624	0.0369	0.0387	0.0479	0.0447	0.0396	0.0291	0.0377	0.0425	0.0376	0.0440
WQ2/20	0.0536	0.0603	0.0405	0.0364	0.0445	0.0336	0.0397	0.0274	0.0365	0.0414	0.0372	0.0409
WQ3/20	0.0554	0.0714	0.0393	0.0371	0.0512	0.0471	0.0494	0.0293	0.0386	0.0415	0.0365	0.0423

Table 4 Combining weight results

Combining W	AQ1	AQ2	G1	G2	G3	G4	G5	G6	G7	G8	G9	G10
WQ1/19	0.05665	0.05070	0.03824	0.03912	0.03446	0.03879	0.03515	0.03993	0.02773	0.03659	0.04054	0.03914
WQ2/19	0.05426	0.04660	0.03925	0.04264	0.03587	0.04122	0.03620	0.03647	0.02812	0.03637	0.04722	0.04202
WQ3/19	0.06258	0.04132	0.03905	0.04308	0.03528	0.04179	0.03550	0.03381	0.03564	0.04131	0.04546	0.04383
WQ4/19	0.05292	0.04292	0.04110	0.04228	0.03479	0.04314	0.03397	0.03895	0.03388	0.03868	0.04384	0.04351
WQ1/20	0.05423	0.04470	0.03810	0.04314	0.03507	0.04111	0.03396	0.03354	0.04092	0.04211	0.03976	0.03836
WQ2/20	0.05537	0.05781	0.04953	0.04015	0.03647	0.04066	0.03293	0.03372	0.03308	0.04129	0.04482	0.04566
WQ3/20	0.05133	0.04738	0.03961	0.03779	0.03461	0.04080	0.03454	0.03292	0.03204	0.03771	0.03981	0.03569

Combining W	LR1	LR2	LR3	LR4	LR5	PR1	PR2	PR3	PR4	PR5	VR1	VR2
WQ1/19	0.05418	0.05916	0.04183	0.04317	0.05003	0.03817	0.04460	0.03101	0.03758	0.04064	0.03786	0.04471
WQ2/19	0.05618	0.05772	0.03963	0.04182	0.04975	0.03363	0.04439	0.02877	0.03807	0.04186	0.03759	0.04432
WQ3/19	0.05308	0.05959	0.03915	0.03943	0.04350	0.03622	0.03824	0.02743	0.03788	0.04262	0.03873	0.04548
WQ4/19	0.05331	0.06224	0.03648	0.04111	0.04594	0.03719	0.03920	0.02976	0.03840	0.04430	0.03864	0.04347
WQ1/20	0.05687	0.06184	0.03600	0.03888	0.04722	0.04513	0.03915	0.02925	0.03737	0.04215	0.03739	0.04375
WQ2/20	0.05341	0.05976	0.03946	0.03658	0.04382	0.03390	0.03924	0.02755	0.03617	0.04105	0.03693	0.04062
WQ3/20	0.05525	0.07082	0.03836	0.03723	0.05040	0.04756	0.04886	0.02944	0.03832	0.04117	0.03631	0.04205

Table 5 Results of TOPSIS in 2019

Number	Bank code	Q1_2019	Q2_2019	Q3_2019	Q4_2019
1	CTG	17	18	22	2
2	BID	19	19	21	18
3	VCB	7	4	5	11
4	MBB	6	10	16	17
5	ACB	11	6	12	14
6	STB	4	13	9	19
7	TCB	8	9	7	3
8	SHB	12	12	15	21
9	EIB	22	21	23	23
10	SCB	13	22	18	1
11	NCB	23	14	8	10
12	KLB	18	11	19	20
13	VPB	14	5	6	8
14	VIB	5	2	2	6
15	HDBank	15	17	14	12
16	SGB	16	3	1	15
17	PGBank	20	23	3	7
18	TPBank	3	8	11	5
19	BAC A Bank	9	20	20	22
20	ABB	21	7	17	9
21	LIEN VIET	10	1	13	13
22	NAM A	1	15	10	16
23	OCB	2	16	4	4

Table 6 demonstrated TOPSIS ranking results of selected credit organizations in 2020, which were figured out under mutual considerations of combining weights. Accordingly, during the nascent stage of the pandemic, ABB, OCB, VPB, TCB, PGBank, VIB, LIEN VIET, KLB, SCB, and TPBank appeared as the best performing banks, whereas, in Q2, 2020 there were several variations in the rankings since SGB, PGBank, VIB, HDBank, TPBank, OCB, Nam A, TCB, VCB entered the top 10. In addition, following this fast-changing trend, standings in the Q3 were once jeopardized with the arrivals of ACB, SHB, and EIB, making the 2020's first three quarters a sensitive period for the banking system. Additionally, since the banks' performances experienced many ups and downs, comparative graphs are thereby applied to demonstrate these changes visually.

For a sensitive analysis, this chapter presents a result of Spearman's correlation (coefficient) values. The quarterly ranking results of each alternative revealed that coefficient values are greater than 0.9, meaning that the two processes have close correlations.

Table 6 Results of TOPSIS in 2020

Number	Bank code	With combining weights			With objective weights		
		Q1_2020	Q2_2020	Q3_2020	Q1_2020	Q2_2020	Q3_2020
1	CTG	19	13	19	19	13	19
2	BID	17	21	18	17	21	18
3	VCB	11	10	13	11	10	13
4	MBB	16	14	12	16	14	12
5	ACB	14	18	4	14	18	4
6	STB	18	17	20	18	17	20
7	TCB	4	7	8	4	7	8
8	SHB	21	16	6	21	16	6
9	EIB	20	23	9	20	23	9
10	SCB	9	15	11	10	15	11
11	NCB	12	22	16	12	22	16
12	KLB	8	11	14	8	11	14
13	VPB	3	4	7	3	5	7
14	VIB	6	3	1	6	3	1
15	HDBank	13	6	2	13	6	2
16	SGB	22	1	23	22	1	22
17	PGBank	5	2	21	5	2	21
18	TPBank	10	8	5	9	8	5
19	BAC A Bank	23	19	22	23	19	23
20	ABB	1	20	15	1	20	15
21	LIEN VIET	7	12	17	7	12	17
22	NAM A	15	9	3	15	9	3
23	OCB	2	5	10	2	4	10

5 Discussions

Today, Vietnam's financial sector is split into two groups, including Commercial Banks and State banks. Accordingly, the State Bank will undertake the responsibility of issuing and managing money. Participate in advisory duties to the Government and State of Vietnam. With policies related to money, such as interest rates, the issue of pre-generation. Currency rate, bank's business draft. Manage foreign currency reserves and credit institutions within the banking system.

Some estimates suggest that the COVID-19 fallout would include decreased demand, lowered wages, an increase in failures, and layoffs in the banking sector [61]. The country is forced to devote all of its available resources to minimize the losses and reduce its adverse effects [62]. It is also further suggested to consider the effect of the model's parts concerning the economic, political, and banking systems to evaluate it more thoroughly because doing so neglects the value of each portion when considering other factors causes it to produce inaccurate results [63]. Furthermore, as studied by [64], their results also indicated that the most important criteria are profitability Ratios. They also used the TOPIS model for the outranking of banks in Turkey's datasets from 2002 to 2011. lack of accurate data, they were left out of the rating model, resulting in improved operating performance.

This chapter has the same direction as [63], which also proposed the TOPSIS method with 35 parameters to rank 15 banks for 2014 to 2018. Due to a lack of data in this study, the components of the Vietnamese bank system, political circumstances, financial instruments, and law observance were among the elements that were omitted. However, this chapter considers the CRITIC weighting method to support the TOPSIS model to have more accurate ranking results. Besides, the experts' judgments on proposed criteria also help this research understand the cause-effect relationship. Overall, the banking sector's operations and growth in 2019 were sound, and its risk management abilities had improved. According to the results in this chapter, it was discovered that the COVID-19 shock's negative effect on corporates was more severe and long-lasting than the other non-bank sectors. Moreover, larger commercial banks, and to a lesser degree better-capitalized banks, also saw lower stock returns, reflecting their more significant anticipated position in coping with the crisis.

As far as DEMATEL weights are concerned, the liquidity ratios play a crucial role in the proposed financial indicators. As is evident from the study, liquidity support and borrower assistance interventions had the most significant positive effect on abnormal bank returns. Illiquid banks have gained most from liquidity support, while large commercial banks have seen a rise in abnormal returns with the announcement of borrower assistance policies. Nonetheless, they are dependent on fiscal expansion and thus have not had a beneficial effect in countries where there is less potential for fiscal expansion haven has already occurred.

Since personal loans are less affected by the disease outbreak than corporate loans, the retail sector is less affected by market cycles. Person and consumer loans were unaffected by the pandemic, but residential and manufacturing loans were severely impacted. Banks almost always lend to companies on the basis of each customer's cash flow condition. The economic impact varies depending on the country and the market structure. Industry—Standard credit from the tertiary sector impacted by the pandemic presents a higher risk to the banks. Transport, warehousing, and postal industries are critical components of state-owned bank lending, while retail and manufacturing are core components of small and medium-sized bank lending. Furthermore, commercial banks cater to small and medium-sized companies amplifying their systemic risk.

Furthermore, liquidity funding for the commercial banking sector is also supported by the central bank [65]. Likewise, Sarı also used the TOPSIS methodology to measure the performance of 11 Turkish banks. However, their approach was based on regression analysis to identify 13 parameters. The results showed that the most relevant criteria were Profitability ratios; Asset quality [66].

6   Conclusions

Firstly, this chapter evaluates the Vietnamese banking sector's performance using a hybrid MCDM approach of CRITIC- DEMATEL and TOPSIS. In this chapter, the objective weight of criteria is obtained by the CRITIC technique. The DEMATEL model is deployed to achieve the subjective weights. Then, 23 commercial banks are evaluated and ranked by TOPSIS. The developed model is useful for determining the position of the banks concerning their competitors. Secondly, this chapter also analyzes COVID-19 impacts on the banking system in Vietnam, basing on the dataset from 2019 to Q3/2020.

From the finding of this chapter, the COVID-19 pandemic has a negative effect on the operations of banks. This is due to the fact that funds that have arrived at the banks on time will not arrive on time, causing a delay. This will make it more difficult for the bank to control operations with these funds prepared ahead of time. As ranking results of 2020, the positions of top banks have been changed. During a downturn, banks face the challenge of generating a supply to meet demand, and banks might have a greater need to seek additional capital to balance the situation.

Regarding the short-term impacts, demand for short-term credit from the residential sector has decreased substantially, while non-financial companies have increased. People were quarantined as a result of the pandemic, and credit card debt plummeted as a result. To save interest charges, borrowers return to work early. Non-performing loans could rise due to the pandemic. If the economy remains stable, bank asset quality will stay the same.

The countries with higher credit and GDP than Vietnam are at greater risk, but their overall effects are minimal. Even so, credit risks rose slightly in the short term. When existing credit risk strains increase, the pandemic lingers on. Financial institutions will concentrate on the refinement and enhancement of the credit system and the development and restructuring of operations.

Concerning medium-term impacts, the financial sector would be impacted by the advent of the next pandemic. The markets have fluctuated due to investor sentiment. The physical economy and the financial markets are currently facing two crises. Pandemic transmissibility makes international crossover and shared transmission more possible. To promote and fund businesses and individuals, State banks have launched numerous initiatives.

References

1. Mosser, P.C.: Central bank responses to COVID-19. Bus. Econ. **55**(4), 191–201 (2020)
2. Zhang, D., Hu, M., Ji, Q.: Financial markets under the global pandemic of COVID-19. Financ. Res. Lett. **36**, 101528 (2020)
3. World Bank: The global economic outlook during the COVID-19 pandemic: a changed world. The World Bank (2020) .

4. Dang, T.T., Wang, C.N., Hiep, N.,Nguyen, N.A.T.: Bank performance evaluation using data envelopment analysis: a case study in Vietnam. In: Contemoorary issues in banking and finance sustainability, fintech and uncertainties university (2020)
5. Tran, T.-T., et al.: Influencing Factors of the International Payment Service Quality at Joint Stock Commercial Bank for Investment and Development of Vietnam. J. Asian Financ. Econ. Bus. **7**(10), 241–254 (2020)
6. Wang, C.N., Luu, Q.C., Nguyen, T.K.L., Der Day, J.: Assessing bank performance using dynamic SBM model. Mathematics (2019)
7. Wu, D.D., Olson, D.L.: Pandemic Risk Management in Operations and Finance: Modeling the Impact of COVID-19 (2020)
8. Nguyen, P.H., Tsai, J.F., Kumar, V.A.G., Hu, Y.C.: Stock investment of agriculture companies in the Vietnam stock exchange market: An AHP integrated with GRA-TOPSIS-MOORA approaches. J. Asian Financ. Econ. Bus. **7**(7), 113–121 (2020)
9. Rodrigues, L., Rodrigues, L.: Economic-financial performance of the Brazilian sugarcane energy industry: An empirical evaluation using financial ratio, cluster and discriminant analysis. Biomass Bioenergy **108**(November), 289–296 (2018)
10. Gudiel Pineda, P.J., Liou, J.J.H., Hsu, C.C., Chuang, Y.C.: An integrated MCDM model for improving airline operational and financial performance. J. Air Transp. Manag. **68**, 103–117 (2018)
11. Haris, M., HongXing, Y., Tariq, G., Malik, A.: An evaluation of performance of public sector financial institutions: Evidence from Pakistan. Int. J. Bus. Perform. Manag. **20**(2), 145–163 (2019)
12. Sharma, A., Jadi, D.M., Ward, D.: Evaluating financial performance of insurance companies using rating transition matrices. J. Econ. Asymmetr. **18**(May), e00102 (2018)
13. Slavica, T.V.: A. Finance, banking and insurance, 1–256 (2017)
14. Beheshtinia, M.A., Omidi, S.: A hybrid MCDM approach for performance evaluation in the banking industry. Kybernetes **46**(8), 1386–1407 (2017)
15. Gasbarro, D., Sadguna, I.G.M., Zumwalt, J.K.: The changing relationship between CAMEL ratings and bank soundness during the Indonesian banking crisis. Rev. Quant. Financ. Account. **19**(3), 247–260 (2002)
16. Roman, A., Şargu, A.C.: Analysing the Financial Soundness of the Commercial Banks in Romania: An Approach based on the Camels Framework. Proc. Econ. Financ. **6**(13), 703–712 (2013)
17. Benjamin, S.J., Bin Mohamed, Z., Marathamuthu, S.: DuPont analysis and dividend policy: empirical evidence from Malaysia Abstract (2016)
18. Rashid, A., Jabeen, S.: Analyzing performance determinants: conventional versus Islamic Banks in Pakistan. Borsa Istanbul Rev. (2016)
19. Bucevska, V., Hadzi Misheva, B.: The determinants of profitability in the banking industry: empirical research on selected Balkan Countries. East. Europ. Econ. **55**(2), 146–167 (2017)
20. Pekkaya, M., Demir, F.E.: Determining the priorities of CAMELS dimensions based on bank performance. Contrib. Econ.:445–463 (2018)
21. Bunea, O.I., Corbos, R.A., Popescu, R.I.: Influence of some financial indicators on return on equity ratio in the Romanian energy sector—a competitive approach using a DuPont-based analysis. Energy **189**, 116251 (2019)
22. Chang, C.T., Ouyang, L.Y., Teng, J.T., Lai, K.K., Cárdenas-Barrón, L.E.: Manufacturer's pricing and lot-sizing decisions for perishable goods under various payment terms by a discounted cash flow analysis Int. J. Prod. Econ. **218**, 83-95 (2019)
23. Le, T.D., Ngo, T.: The determinants of bank profitability: A cross-country analysis. Cent. Bank Rev. **20**(2), 65–73 (2020)
24. Carras, M.A., Knowler, D., Pearce, C.M., Hamer, A., Chopin, T., Weaire, T.: A discounted cash-flow analysis of salmon monoculture and Integrated Multi-Trophic Aquaculture in eastern Canada. Aquac. Econ. Manag. **24**(1), 43–63 (2020)
25. Wu, H.Y.: Constructing a strategy map for banking institutions with key performance indicators of the balanced scorecard. Eval. Program Plann. **35**(3), 303–320 (2012)

26. Park, J.H., Shea, C.H., Wright, D.L.: Reduced-frequency concurrent and terminal feedback: a test of the guidance hypothesis. J. Mot. Behav. (2000)
27. Schalock, R.L., Bonham, G.S.: Measuring outcomes and managing for results. Eval. Program Plann. **26**(3), 229–235 (2003)
28. Sridharan, S., Go, S., Zinzow, H., Gray, A., Barrett, M.G.: Analysis of strategic plans to assess planning for sustainability of comprehensive community initiatives. Eval. Program Plann. **30**(1), 105–113 (2007)
29. Liu, J.S., Lu, L.Y.Y., Lu, W.M., Lin, B.J.Y.: A survey of DEA applications. Omega (United Kingdom) (2013)
30. Cooper, W.W., Seiford, L.M., Zhu, J.: Data envelopment analysis: history, models, and interpretations. Int. Ser. Oper. Res. Manag. Sci. (2011)
31. Gregoriou, G.N.: Quantitative models for performance evaluation and benchmarking: data envelopment analysis with spreadsheets. J. Wealth Manag. **17**(4), 114–115 (2015)
32. Paradi, J.C., Zhu, H.: A survey on bank branch efficiency and performance research with data envelopment analysis. Omega (United Kingdom) (2013)
33. Wang, K., Huang, W., Wu, J., Liu, Y.N.: Efficiency measures of the Chinese commercial banking system using an additive two-stage DEA. Omega (United Kingdom) (2014)
34. Wanke, P., Barros, C.: Two-stage DEA: An application to major Brazilian banks. Expert Syst. Appl. (2014)
35. Othman, F.M., Mohd-Zamil, N.A., Rasid, S.Z.A., Vakilbashi, A., Mokhber, M.: Data envelopment analysis: a tool of measuring efficiency in banking sector," Int. J. Econ. Financ. (2016)
36. Akkoç, S., Vatansever, K.: Fuzzy performance evaluation with AHP and topsis methods: evidence from Turkish Banking Sector after the global financial crisis. Eurasian J. Bus. Econ. **6**(11), 53–74 (2013)
37. Nguyen, P.: A Fuzzy Analytic Hierarchy Process (FAHP) Based on SERVQUAL for Hotel Service Quality Management : Evidence from Vietnam *. J. Asian Financ. Econ. Bus. **8**(2), 1101–1109 (2021)
38. Nguyen, P.-H., Tsai, J.-F., Nguyen, T.-T., Nguyen, T.-G., Vu, D.-D.: A Grey MCDM Based on DEMATEL Model for Real Estate Evaluation and Selection Problems: A Numerical Example. J. Asian Financ. Econ. Bus. **7**(11), 549–556 (2020)
39. Nguyen, P.H., Tsai, J.F., Nguyen, H.P., Nguyen, V.T., Dao, T.K.: Assessing the Unemployment Problem Using A Grey MCDM Model under COVID-19 Impacts: A Case Analysis from Vietnam. J. Asian Financ. Econ. Bus. **7**(12), 53–62 (2020)
40. Nguyen, P.H., Tsai, J.F., Nguyen, V.T., Vu, D.D., Dao, T.K.: A Decision Support Model for Financial Performance Evaluation of Listed Companies in The Vietnamese Retailing Industry. J. Asian Financ. Econ. Bus. **7**(12), 1005–1015 (2020)
41. Dinçer, H., Yüksel, S.: Comparative Evaluation of BSC-Based New Service Development Competencies in Turkish Banking Sector with the Integrated Fuzzy Hybrid MCDM Using Content Analysis. Int. J. Fuzzy Syst. **20**(8), 2497–2516 (2018)
42. Zhao, Q., Tsai, P.H., Wang, J.L.: Improving financial service innovation strategies for enhancing China's banking industry competitive advantage during the fintech revolution: A hybrid MCDM model. Sustain. **11**(5), 1–29 (2019)
43. Wu, H.Y., Tzeng, G.H., Chen, Y.H.: A fuzzy MCDM approach for evaluating banking performance based on balanced scorecard. Expert Syst. Appl. (2009)
44. Nassereddine, M., Eskandari, H.: An integrated MCDM approach to evaluate public transportation systems in Tehran. Transp. Res. Part A Policy Pract. **106**(April), 427–439 (2017)
45. Pak, J.Y., Thai, V.V., Yeo, G.T.: Fuzzy MCDM Approach for Evaluating Intangible Resources Affecting Port Service Quality. Asian J. Shipp. Logist. **31**(4), 459–468 (2015)
46. Akdag, H., Kalayci, T., Karagöz, S., Zülfikar, H., Giz, D.: The evaluation of hospital service quality by fuzzy MCDM. Appl. Soft Comput. J. **23**, 239–248 (2014)
47. Zoraghi, N., Amiri, M., Talebi, G., Zowghi, M.: A fuzzy MCDM model with objective and subjective weights for evaluating service quality in hotel industries. J. Ind. Eng. Int. **9**(1), 1–13 (2013)

48. Tsaura, S.H., Chang, T.Y., Yen, C.H.: The evaluation of airline service quality by fuzzy MCDM. Tour. Manag. **23**(2), 107–115 (2002)
49. Tseng, M.L.: Using hybrid MCDM to evaluate the service quality expectation in linguistic preference. Appl. Soft Comput. J. **11**(8), 4551–4562 (2011)
50. Si, S.L., You, X.Y., Liu, H.C., Zhang, P.: DEMATEL Technique: A Systematic Review of the State-of-the-Art Literature on Methodologies and Applications. Math. Probl. Eng. **1**, 2018 (2018)
51. Hsu, C.W., Kuo, T.C., Chen, S.H., Hu, A.H.: Using DEMATEL to develop a carbon management model of supplier selection in green supply chain management. J. Clean. Prod. (2013)
52. Asan, U., Kadaifci, C., Bozdag, E., Soyer, A., Serdarasan, S.: A new approach to DEMATEL based on interval-valued hesitant fuzzy sets. Appl. Soft Comput. J. (2018)
53. Govindan, K., Chaudhuri, A.: Interrelationships of risks faced by third party logistics service providers: a DEMATEL based approach. Transp. Res. Part E Logist. Transp. Rev. (2016)
54. Malviya, R.K., Kant, R.: Hybrid decision making approach to predict and measure the success possibility of green supply chain management implementation. J. Clean. Prod. (2016)
55. Patil, S.K., Kant, R.: A hybrid approach based on fuzzy DEMATEL and FMCDM to predict success of knowledge management adoption in supply chain. Appl. Soft Comput. J. (2014)
56. Rostamzadeh, R., Ghorabaee, M.K., Govindan, K., Esmaeili, A., Nobar, H.B.K.: Evaluation of sustainable supply chain risk management using an integrated fuzzy TOPSIS- CRITIC approach. J. Clean. Prod. (2018)
57. Abdel-Basset, M., Mohamed, R.: A novel plithogenic TOPSIS- CRITIC model for sustainable supply chain risk management. J. Clean. Prod. (2020).
58. Keshavarz Ghorabaee, M., Amiri, M., Kazimieras Zavadskas, E., Antuchevičienė, J.: Assessment of third-party logistics providers using a CRITIC–WASPAS approach with interval type-2 fuzzy sets. Transport (2017)
59. Criteria, T., Through, I., Correlation, I.: CRITIC Method, pp 5–7 (1995)
60. Gabus, A., Fontela, E.: World problems an invitation to further thought within the framework of DEMATEL. Battelle Geneva Res. Cent., 1–8 (1972)
61. Kimbonguila, A., Matos, L., Petit, J., Scher, J., Nzikou, J.M.: Effect of physical treatment on the physicochemical, rheological and functional properties of Yam Meal of the cultivar 'Ngumvu' from dioscorea Alata L. of Congo. Int. J. Recent Sci. Res. (2019)
62. Mamadiyarov, Z., Azlarova, A.: Covid 19 visits to banking institutions—yesterday , today and tomorrow (2021)
63. Abbasi, S., Nazemi, A.: Presenting and evaluating the banks rating model using topsis technique. Int. J. Nonlinear Anal. Appl. **11**, 195–209 (2020)
64. Önder, E., Hepsen, A.: Combining time series analysis and multi criteria decision making techniques for forecasting financial performance of banks in Turkey. Int. J. Latest Trends Financ. Econ. Sci. (2013)
65. Heffernan, T., Pawlak, M.: Crisis futures: The affects and temporalities of economic collapse in Iceland. Hist. Anthropol. Chur. **31**(3), 314–330 (2020)
66. Sarı, T.: Performance evaluation of Turkish banks with TOPSIS and stepwise regression (2020)
67. Tanasković, S., Jandrić, M.: Macroeconomic and institutional determinants of non-performing loans. J. Cent. Bank. Theory Pract. (2015)
68. Makri, V., Tsagkanos, A., Bellas, A.: Determinants of non-performing loans: the case of Eurozone. Panoeconomicus (2014)
69. Lipson, M.L., Mortal, S., Schill, M.J.: On the scope and drivers of the asset growth effect. J. Financ. Quant. Anal., 1651–1682 (2011)
70. Juárez, F.: The growth of companies as a function of total assets. WSEAS Trans. Bus. Econ. **15**, 301–310 (2018)
71. Jo, H., Han, I., Lee, H.: Bankruptcy prediction using case-based reasoning, neural networks, and discriminant analysis. Expert Syst. Appl. **13**(2), 97–108 (1997)
72. Watson, J.: Comparing the performance of male-and female-controlled businesses: relating outputs to inputs. Entrep. theory Pract. **26**(3), 91–100 (2002)

73. Fernández de Lis, S., Martínez Pagés, J., Saurina Salas, J.: Credit growth, problem loans and credit risk provisioning in Spain. Banco de España. Servicio de Estudios (2000)
74. De Lis, F.S., Pagés, J.M., Saurina, J.: Credit growth, problem loans and credit risk provisioning in Spain. BIS Pap. **1**, 331–353 (2001)
75. Hati, S.R.H., Wibowo, S.S., Safira, A.: The antecedents of Muslim customers' intention to invest in an Islamic bank's term deposits: evidence from a Muslim majority country. J. Islam. Mark. (2020)
76. Yulianto, A., Solikhah, B.: The internal factors of Indonesian Sharia banking to predict the mudharabah deposits. Rev. Integr. Bus. Econ. Res. **5**(1), 210 (2016)
77. Duguma, G.J., Han, J.: Effect of deposit mobilization on the financial sustainability of rural saving and credit cooperatives: evidence from Ethiopia. Sustainability **10**(10), 3387 (2018)
78. Tuyishime, R., Memba, F., Mbera, Z.: The effects of deposits mobilization on financial performance in commercial banks in Rwanda: a case of equity bank Rwanda limited. Int. J. small Bus. Entrep. Res. **3**(6), 44–71 (2015)
79. Li, L., Zhang, Y.: Are there diversification benefits of increasing noninterest income in the Chinese banking industry? J. Empir. Financ. **24**, 151–165 (2013)
80. Maudos, J., Solís, L.: The determinants of net interest income in the Mexican banking system: an integrated model. J. Bank. Financ. **33**(10), 1920–1931 (2009)
81. Vozková, K., Teplý, P.: An analysis of bank fee and commission income in the EU and in the Czech Republic in a low interest rate environment. Sci. Pap. Univ. Pardubice. Ser. D, Fac. Econ. Adm. **28**(2) (2020)
82. Köhler, M.: An analysis of non-traditional activities at German savings banks: Does the type of fee and commission income matter? (2018)
83. UYÊN, T. T. Ú.: The impact of initial public offering on profit before tax on asset in vietnamese enterprises–from the perspective of management accounting. J. Econ. Dev. 50–57 (2019)
84. Steffens, P., Davidsson, P., Fitzsimmons, J.: Performance configurations over time: implications for growth–and profit–oriented strategies. Entrep. Theory Pract. **33**(1), 125–148 (2009)
85. Nariswari, T.N., Nugraha, N.M.: Profit growth: impact of net profit margin, gross profit margin and total assests turnover. Int. J. Financ. Bank. Stud. **9**(4), 87–96 (2020)
86. Scherer, F.M.: Corporate inventive output, profits, and growth. J. Polit. Econ. **73**(3), 290–297 (1965)
87. Cebenoyan, A.S., Strahan, P.E.: Risk management, capital structure and lending at banks. J. Bank. Financ. (2004)
88. Messai, A.S., Jouini, F.: Micro and macro determinants of non-performing loans. Int. J. Econ. Financ. (2013)
89. Nissim, D., Penman, S.H.: Ratio analysis and equity valuation: from research to practice. Rev. Account. Stud. (2001)
90. Heikal, M., Khaddafi, M., Ummah, A.: Influence analysis of return on assets (ROA), return on equity (ROE), net profit margin (NPM), debt to equity ratio (DER), and current ratio (CR), against corporate profit growth in automotive in indonesia stock exchange. Int. J. Acad. Res. Bus. Soc. Sci. (2014)
91. Mawaddah, N.: Faktor-Faktor Yang Mempengaruhi Profitabilitas Bank Syariah. ETIKONOMI (2015)
92. Busch, R., Memmel, C.: Banks' net interest margin and the level of interest rates. Credit Cap. Mark. (2017)
93. Rosly, S.A., Abu Bakar, M.A.: Performance of Islamic and mainstream banks in Malaysia. Int. J. Soc. Econ. (2003)
94. Robin, I., Salim, R., Bloch, H.: Financial performance of commercial banks in the post-reform era: further evidence from Bangladesh. Econ. Anal. Policy (2018)
95. Petria, N., Capraru, B., Ihnatov, I.: Determinants of banks' profitability: evidence from EU 27 banking systems. Proc. Econ. Financ. (2015)
96. Hirtle, B.J., Stiroh, K.J.: The return to retail and the performance of US banks. J. Bank. Financ. (2007)

97. Lo, C.W., Leow, C.S.: Islamic banking in Malaysia: a sustainable growth of the consumer market. Int. J. Trade Econ. Financ. (2014)
98. Zhang, H.: Share price performance following actual share repurchases. J. Bank. Financ. (2005)
99. Agostino, M., Drago, D., Silipo, D.B.: The value relevance of IFRS in the European banking industry. Rev. Quant. Financ. Account. (2011)

Policy Response to Covid-19 Pandemic and Its Impact on the Vietnamese Economy: An Analysis of Social Media

Lan T. M. Nguyen and Soan T.M. Duong

Abstract Amid the outburst of the Covid-19 pandemic worldwide, many countries are hesitant to implement aggressive measures to curb the spread of the pandemic with the fear that this would damage their national economy. A question is whether there is an actual trade-off between saving lives and economic development. To answer this, we study the case of Vietnam, a successful story in combating the pandemic thanks to the Government's immediate and rigorous policies. We collect official news related to the Covid-19 pandemic in Vietnam in the year 2020 by utilizing a self-developed web scraping tool. By analyzing the news data, we identify the key phases of the pandemic in Vietnam and the key Government policy announcements. We then contrast this information with the performance of the equity market and find that the two types of policies, including border and entry control and cross-country travel control, negatively impact the equity market, whilst the other policies hardly affect the stock market return. These findings suggest that government policies can be very efficient in controlling the Covid-19 pandemic; however, some of them almost cause no damage to the country's economy. This study, therefore, efficiently depicts the relationship between pandemic combat and economic development. It also provides valuable guidance for other countries in terms of policy decisions during the current fight against the Covid-19 pandemic as well as in future community. health crises.

Keywords Covid-19 · Policy response · Economy development · Social media

L. T. M. Nguyen (✉)
FPT University, Hanoi, Vietnam
e-mail: lanntm3@fe.edu.vn; lanntm.hrc@gmail.com

S. T.M. Duong
Le Quy Don Technical University, Hanoi, Vietnam
e-mail: soanduong@lqdtu.edu.vn

© Institute of Technology PETRONAS Sdn Bhd 2022
S. A. Abdul Karim (eds.), *Shifting Economic, Financial and Banking Paradigm*,
Studies in Systems, Decision and Control 382,
https://doi.org/10.1007/978-3-030-79610-5_3

1 Introduction

The Covid-19 pandemic is a serious public health crisis of humankind that already led to an economic crisis. Some countries were hesitant to implement stringent measures to curb the spread of the pandemic with the fear that this would damage their national economy. In contrast, some decided to take very strict measures from the early stage of the to quickly control the disease, which subsequently stabilizes the economy. Recent literature related to the Covid-19 pandemic mostly shares the focus on the impact of government intervention on controlling the pandemic [1, 2]. Whether such intervention can really dampen the economic situation is a question that is left unanswered.

In this chapter, we focus on examining how government policies to fight the pandemic affect the stock market movement. We employ the stock market as the subject of our research as it has long been considered an efficient indicator and predictor for the economic performance of a country. Beaudry and Portier suggest that stock prices reflect public expectations about the economic conditions [3]. According to Hamilton and Lin, a substantial drop in stock price can predict a future economic downturn, whilst a large increase in stock prices would suggest future growth of the economy [4].

For our empirical analyses, we study the case of Vietnam, a successful country in the fight against the Covid-19 pandemic thanks to the Government's immediate and rigorous intervention policies. As of 31 December 2021, the country had reported only 1465 cases of inflection and 35 death of Covid-19, of which most of the inflected cases are from overseas immigrants [5]. Regarding the governmental responses to the outbreak of the pandemic, the Vietnamese Government was among the few countries that closed its borders, shut its schools and halt international flights during the very early stage of the pandemic [6]. By late March 2020, Vietnam had already employed the partial then full social distancing with several cultural and social events being canceled and domestic traveling being reduced by 60 to 70% compared to March 2019 [7]. Although Vietnam's strict measures to combat the pandemic have negatively affected businesses nationwide, the Government also provides multiple supports and stimulus packages via its fiscal and monetary policies to stabilize the social and economic situations. Vietnam is among a rare list of countries that records positive economic growth during the pandemic [8]. It is reported that some key indicators, including retail sales, imports, and industrial production of Vietnam in June 2020, are up compared to the same period in 2019, and Vietnam is forecasted to be one of the fastest-growing economies in Southeast Asia [9]. Altogether, Vietnam is an interesting case to study how governmental policies affect economic performance. The outcome of this study can also provide a useful lesson for other countries while constructing their own policies to combat the current worldwide pandemic.

The main objectives of this study are as follows:

(a) To identify key Government policies of Vietnam during the Covid-19 pandemic,
(b) To identify key phases of the Covid-19 pandemic in Vietnam,

(c) To evaluate the impact that different types of Government policies in different phases of the pandemic have on the Vietnamese stock market.

This chapter is organized as follows. In Sect. 1, we provide a brief introduction as well as a related literature review. Section 2 is an overview of the Covid-19 pandemic and its impact on the global economy. Section 3 describes the data collection process and the methodology. Sections 4 and 5 are the discussion of the key phases of the pandemic in Vietnam and the Government responses to control the pandemic, respectively. Section 6 is dedicated to the results and discussion about the impact of the government policies on the Vietnamese stock market. Section 7 concludes our work.

2   Overview of Covid-19 Pandemic and Government Policy Response

The Covid-19 pandemic is so far the most serious public health crisis that also triggered an economic crisis globally. Compared to the global financial crisis (GFC) in 2007–2008, this current crisis also affects all countries in the world; however, the degree of economic damage is more severe. Tooze shows that during the GFC, the global GDP fell by 2.1%, whereas as of Quarter 2 of 2020, the Covid-19 pandemic had caused a decrease of 5.5% in global GDP [10].

The reason for such severe impact on the world economy caused by the Covid-19 pandemic lies in its nature. The GFC was triggered by the escalating credit risks that had piled up in the financial sectors then spilled over to the other sectors of the economy. The Covid-19 crisis results from an unexpected pandemic that causes an immediate impact on all sectors. Governments worldwide have to implement rigorous measures to curb the pandemic, including social distancing, school closures, and border shutdown, putting the freeze on the whole economy.

In addition, the degree of government intervention during the two crises is another reason for the different impact of the crises on the economy. During the GFC in 2007–2008, the focus of government intervention worldwide is directed to supporting economic recovery, which involves several fiscal and monetary policies [11]. On the contrary, government policies during the Covid-19 crisis involve several actions in terms of social, medical, educational, and financial measures [2]. On the one hand, such extensive intervention is efficient in controlling the pandemic. On the other hand, it can cause severe damage to the national economy that might not be reversed in the foreseeable future.

3   Research Design and Data Collection

The aim of this chapter is to analyze the policy response to the Covid-19 pandemic in Vietnam and its impact on the economy. To achieve that, we first identified the Covid phases in Vietnam in 2020 via the daily traffic of Covid-19 related news on the social media. We then manually searched for the policies related to the Covid-19 pandemic as indicated by the news pattern. Finally, we introduced a method to evaluate the impact of those Government policies on the stock market in Vietnam.

3.1   Social Media Mining

To identify the phases of the Covid-19 pandemic based on social media data, we first searched for the daily traffic of news related to Covid-19 within the studied period; this is inspired by the work in [12]. To be more precise, we found articles that include Covid-19 related keywords, i.e. *coronavirus, corona virus, covid, ncov, sars-cov*, and *viem phoi* (Vietnamese for pneumonia), from the top-ten common online newspapers in Vietnam. According to the statistic of Amazon Alexa, the top-ten online newspapers in Vietnam are: (1) *vnexpress.net*, (2) *laodong.vn*, (3) *24h.com.vn*, (4) *zingnews.vn*, (5) *kenh14.vn*, (6) *dantri.com.vn*, (7) *tuoitre.vn*, (8) *vietnamnet.vn*, (9) *cafef.vn*, and (10) *thanhnien.vn* [13].

To search for the Covid-19 related articles automatically, we built a Python toolbox for crawling articles on the ten selected websites. In detail, for each newspaper site, we developed a crawler class, which is inherited `CrawlSpider` class from Scrapy engine [14]. The main idea of a crawler class is that it sets a loop to visit all article links of a predefined domain. The articles which were published in 2020 and includes Covid-19 related keywords were then recorded.

We next counted the number of articles per day using the recorded articles. We then separated the period of interest into phases by identifying the local increasing and decreasing of Covid-19 related news. In detail, we smoothed the curve of daily news numbers, computed the gradient of the smoothness curve, and considered the local peaks and troughs as the separated points. Finally, we focus on the local peaks of news from our social media analyses and manually search for an announcement of Government policies or a special infected case or death on the day of peak news. The toolbox and extracted data are available at: https://github.com/soanduong/Covid-news-crawlers.

3.2   Data Analyses

In this study, we employ the event study methodology for our analyses of stock market responses to the Vietnamese Government policies. We focus on the VNINDEX as the

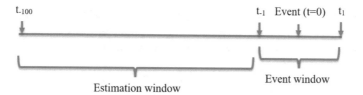

Fig. 1 Timeline of the proposed event study

indicator for the economic performance of Vietnam. The historical data of VNINDEX is retrieved from *cafef.vn*, a prestigious website containing financial news and data of Vietnamese firms.

From our social media mining, we are able to identify key events related to the Covid-19 pandemic in Vietnam, including special announced cases/deaths and key Government policies. Accordingly, we identify the event window and estimation window as followed. The event date is captured at $t = 0$. The event window is the period surrounding the event date. Since the frequency of events in our sample is very high, we choose to only examine the impact of government policy on the stock market on the event date ($t = 0$) alone to only capture the immediate market response to Covid-19 news. This also allows us to avoid the overlapping effect of multiple news on the stock index. We identify our estimation window as 100 trading days prior to our first event date ($t = [-100, -1]$). The timeline for our event study is depicted in Fig. 1.

For our estimation model, we first calculate the daily return of the stock index using Eq. (1):

$$R_t = \ln(P_t/P_{t-1}) * 100, \tag{1}$$

where R_t is the log return of the VNINDEX on day t. P_t and P_{t-1} are the closing price of the index on day t and $t - 1$, respectively.

We next estimate the normal return (i.e., the expected return for VNINDEX should there be no Covid-19 related event). We follow the method in [15], and calculate the normal return (*NR*) as the average return of the index during our estimation window of 100 trading days prior to the first Covid-19 event in Vietnam:

$$NR = \frac{1}{N} \sum_{-100}^{-1} R_t. \tag{2}$$

We then can calculate the abnormal return (*AR*) for the index on each trading day as in Eq. (3)

$$AR_t = R_t - NR. \tag{3}$$

We examine the abnormal return on each trading day that has an announcement of a governmental policy related to the Covid-19 pandemic for the univariate analyses. In addition, based on our social media analyses, we also classify the policies into different types and identify the different phases of the pandemic in Vietnam. We are then able to examine the impact of each type of policy on the stock market AR and to evaluate the impact of governmental policies on the stock market AR in different phases of the pandemic.

For our regression model, we employ the ARMAX model, which is ARMA Model Including Exogenous Covariates. This method is popularly adopted by previous studies to efficiently capture stock market movement [16]. Since our event window is narrow, we decide to employ the ARMAX(1,1) model for our analyses to place more focus on the impact of Covid-19 news on the stock market return.

Our regression model is structured as

$$R_t = \alpha R_{t-1} + \beta \epsilon_{t-1} + \gamma X_t + \epsilon_t, \tag{4}$$

where R_t and R_{t-1} are the return of the stock market index on day t and $t-1$, respectively. ϵ_{t-1} is the moving average term. ϵ_t captures the white noise of the model. And X_t is a vector of exogenous variables, which include *Policy*, *Phase*, *Cases*, *Deaths*, and *Special cases*. *Policy* is a dummy variable that equals 1 if there is an announcement of a Government policy related to the Covid-19 pandemic on day t and zero otherwise. We also examine the impact of different types of policies on the stock market return by utilizing a series of dummy variables, P_i, which represent seven types of policies as per our policy classification. We include four control variable in our model. *Phase* is a series of dummy variables that represent the different phases of the pandemic in Vietnam. *Special cases* is a dummy variable that equals 1 is there is an announcement of a special infected case or death on day t and zero otherwise. *Cases* and *Deaths* represent the total number of infected cases and death as of day t, respectively.

4   Key Phases of the Pandemic in Vietnam

The developed toolbox collected 156,289 Covid-19 related news in 2020 from the ten newspaper sites. Figure 2a shows the number of daily Covid-19 related news in 2020; Fig. 2b shows the gradient of the smoothness daily news curve. From Fig. 2b, we found that the first dramatically increasing Covid-related news is on 23 January on when the first two cases were confirmed in Vietnam [17]. Thus, we identified the first phase from 23 January, 2020. As can be seen in Fig. 2b, there is one local peak of Covid-19 related-news traffic, i.e. in 23 July. There are two local troughs in 23 April and 27 August. Therefore, we identified four phases of the Covid-19 pandemic in 2020.

- **Phase 1** from 23/01/2020 to 23/04/2020. This period is identified as the first Covid-19 wave in Vietnam.
- **Phase 2** from 24/04/2020 to 23/07/2020. This period is approximately normal as there was few comunity transmission Covid-19 cases.

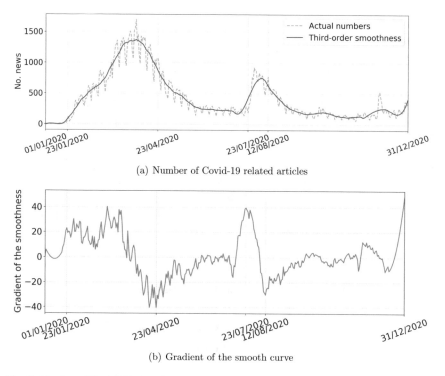

(a) Number of Covid-19 related articles

(b) Gradient of the smooth curve

Fig. 2 Number of Covid-19 related articles per day from the top-ten newspaper websites in 2020. The key dates in the period are marked in the timeline

- **Phase 3** from 24/07/2020 to 12/08/2020. This period is recognized as the second Covid-19 wave of Vietnam.
- **Phase 4** from 28/08/2020 to 31/12/2020. This period is approximately normal.

5   Government Intervention in Vietnam

We focus on the local peaks of news from our social media analyses and manually search for an announcement of Government policies or a special infected case or death on the day of locally peaked news. In total, we find 61 policy announcements and 5 special case/death announcements, which spread over 49 event dates. We further classified Government policies into seven types as followed.

- P_1 (Outbreak announcements and emergency measures): includes announcements about the emergence of new outbreaks of the pandemic, guidelines for emergency situations issued by different levels of Government, and emergency responses by hospitals.

- P_2 (Medical measures): includes the announcements of required medical declarations for immigrants and voluntary to required medical declarations for different groups of citizens.
- P_3 (School closure): includes the announcements of school closure across the countries or in some provinces nationwide.
- P_4 (Border and entry control): includes border shutdown and immigration control announcements.
- P_5 (Social isolation): includes the announcements of required closures of certain types of businesses and social distancing at different levels.
- P_6 (Financial supports): includes the announcements of financial aids from the Government and Government control of prices and commodity pile-up.
- P_7 (Cross-country travel control): includes the announcements of restricted movement between provinces, restricted travel from epidemic zones, reduction or cancellation of domestic flights.

6   Policy Impact on the Vietnamese Economy

6.1   Overview of the Vietnamese Stock Market Under the Impact of the Covid-19 Pandemic

We utilize the historical data of VNINDEX to plot a graph illustrating the movement of the Vietnamese stock market as in Fig. 3. Accordingly, during the year 2020, the VNINDEX witnessed a sharp decrease from around 960 points on 30 January 2020 to its historic low of 662 points at the end of March. During the first three months of 2020, the VNINDEX reduced by 28%, equivalent to a loss of USD 37.4 billion in market capitalization. This reduction equaled over 15% of the gross domestic products (GDP) of Vietnam in 2019. This reflects the negative view of the market during the first wave of the pandemic when all market participants grew skeptical about the pandemic situation and government intervention's effectiveness to curb the widespread of the virus. This decreasing trend ended as the Vietnamese Government issue Decree 16 on 31 March 2020 that imposed the highest level of social distancing across the whole country.

After such a dramatic fall, VNINDEX saw a continuous increase in value until it reached a peak of 900s point on 10 Jun 2020. This increasing trend reflects the positive expectation of all market participants about the outlook of the Vietnamese economy as they have more confidence in the immediate and effective government intervention to control the pandemic. After 10 Jun, the increasing trend turned into a slight fluctuation until the VNINDEX reached its new low of 785 on 27 July. This coincides with the start of the second wave of the Covid-19 pandemic in Vietnam, with the announcement of a new infected case of Covid-19 in the society in Danang. However, we observe that market reaction to the second wave was less severe, with the VNINDEX reducing by only 13%. In addition, this second fall in the VNINDEX

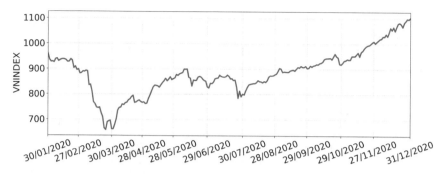

Fig. 3 The performance of VNINDEX in 2020

was soon reversed by a more persistent increasing trend that reached a peak of around 1100 points. This far exceeded the peak in 2019 and almost reached to historic peak in 2018. This shows the high confidence of market participants in the Vietnamese economic conditions whilst most countries in the world were still struggling to control the non-stop spread of the Covid-19 pandemic.

6.2 Univariate Tests for the Impact of Government Intervention Policies on the Vietnamese Stock Market During the Covid-19 Pandemic

We first perform a univariates test to examine the abnormal return (*AR*) of the VNIN-DEX on the event dates. First, Table 1 documents the average *AR* for the days with the announcement of special infected cases and the announcement of a government intervention policy. We find that, on average, the announcement of a special infected case caused a reduction of 2% in the stock market daily return, significant at the one percent level. However, the announcement of government policy shows no significant impact on the stock market. We further examine the days in which a new policy announcement does not coincide with a special case announcement to avoid the overlapping effects of two types of announcements. The results remain consistent in the sense that a policy announcement, on average, does not exert a significant impact on the stock market return.

Next, in Table 2, we examine the impact of Government policies on the stock market during different phases of the pandemic. The results show that Government intervention exerts a negative and significant impact on the stock market during the second phase, while the impact documented in the remaining phases is insignificant. Specifically, an announcement of a new policy during the second phase of the pandemic causes a decline of 1.4% in the daily return of the VNINDEX. The possible explanation for such finding can be because the second phase of the pandemic in Vietnam coincides with the serious outbreak of the pandemic in most countries in

Table 1 Univariate tests for the impact of government policies and the announcement of special Covid-19 cases on the VNINDEX

	AR (%)	T-Value	Significant	Observation
Special cases	−2.0451	−3.81906	***	5
Policy	−0.6215	−1.16050		49
Policy with no special cases	−0.4597	−0.85839		44

This table provides univariate test results for the relationship between government policies and VNINDEX return. *AR* is the daily abnormal return of VNINDEX. *Policy* is a dummy variable that equals 1 if there is an announcement of a governmental policy related to the Covid-19 pandemic on day *t* and zero otherwise. *Special case* is a dummy variable that equals 1 is there is an announcement of a special infected case or death on day *t* and zero otherwise. ***, **, and * represent significance levels of 1%, 5%, and 10%, respectively, for the two-tailed T-test

Table 2 Univariate tests for the impact of government policies by phases on VNINDEX

Phases	AR (%)	T-Value	Significant	Observation
Phase 1 the first Covid wave	−0.7966	−1.48760		33
Phase 2 approximately normal period	−1.4004	−2.61498	***	3
Phase 3 the second Covid wave	0.0813	0.15174		7
Phase 4 approximately normal period	−0.0884	−0.16514		3

This table provides univariate test results for the relationship between government policies and VNINDEX return. AR is the daily abnormal return of VNINDEX. ***, **, and * represent significance levels of 1%, 5%, and 10%, respectively, for the two-tailed T-test

the world. Thus, any policies issued during this time can cause a panic response by the stock market, resulting in a negative reflection in the VNINDEX movement.

In Table 3, we further investigate the impact of government policies on the stock market by policy types. Our results show that Policy type 1 (Outbreak announcements & emergency measures), 4 (Border and entry control), and 7 (Cross-country travel control) show a negative and significant impact on the stock market. This is reasonable since tourism and hospitality services are among the leading sectors of the Vietnamese economy [7]. As such, the control over overseas and domestic travels can significantly affect the country's gross income, which is translated into a decline in the VNINDEX. Specifically, an announcement of a border and entry control policy results in a decline of 2.06% in the VNINDEX return, whereas an announcement of cross-country travel control leads to a reduction of 2% in market return. We also document that policies related to the announcement of pandemic outbreaks and emergency measures (P_1)

Table 3 Univariate tests for the impact of government policies by types on the VNINDEX

Types	AR (%)	T-Value	Significant	Observation
P_1	−1.0593	−1.97817	*	14
P_2	−0.4828	−0.90162		10
P_3	−0.2796	−0.52210		4
P_4	−2.0588	−3.84460	***	9
P_5	−0.3527	−0.65860		9
P_6	−0.7530	−1.40614		11
P_7	−2.0024	−3.73922	***	4

This table provides univariate test results for the relationship between government policies and VNINDEX return. AR is the daily abnormal return of VNINDEX. P_1 through P_7 is dummy variables that represent different types of government policies related to Covid-19 pandemic on day t, the coding of policy types can be found in Sect. 5. ***, **, and * represent significance levels of 1%, 5%, and 10%, respectively, for the two-tailed T-test

result in a 1.06% reduction in the VNINDEX return, significant at the ten percent level.

In contrast, other policies including medical measures (P_2), school closures (P_3) and social isolation (P_5) exert no significant impact on the stock markets. This suggests that although these policies were proved to be very efficient in controlling the Covid-19 pandemic, they almost cause no damage to the country's economy. To this end, we conclude that there is not always a trade-off between saving lives and economic growth.

Surprisingly, we find that the financial supports from the Government (P_6) shows no significant effect on the stock market return. This highlights the fact that Government supports appear to be ineffective in elevating economic growth. This finding is important as it raises the question of how the Government should design its supporting package to facilitate economic benefits.

6.3 Regression Results for the Impact of Government Intervention Policies on the Vietnamese Stock Market During the Covid-19 Pandemic

One empirical issue with our univariate tests is that there can be a few days that have the announcements of more than one policy and/or a special case. The inclusion of those days in our analyses can result in incorrect conclusions. It is also not efficient to remove all those days with overlapping events since it would result in a loss of important information. Thus, in this section, we perform regression analyses utilizing the ARMAX model to overcome this issue.

In Table 4, we present the results of the regression with Policy being our variable of interest. The first regression includes all control variables, and the second

Table 4 Regression results for the impact of government policies on the VNINDEX

Variables	(1) Return (std.)	(2) Return (std.)
Policy	−0.5092**	−0.4514*
	(0.2331)	(0.2653)
Special cases	−1.5178***	−1.3277**
	(0.4643)	(0.5239)
Cases	0.0008	0.0017*
	(0.0006)	(0.0010)
Death	−0.0154	−0.0145
	(0.0182)	(0.0318)
AR(1)	−0.1326	−0.1348
	(2.7370)	(4.1537)
MA(1)	0.1114	0.1204
	(2.7657)	(4.1870)
Sigma	1.4129***	1.4068***
	(0.0457)	(0.0474)
Constant	−0.1421	−1.3734
	(0.2478)	(1.3572)
Observations	237	237
Phase fixed effects	No	Yes

This table provides regression results for the relationship between government policies and VNIN-DEX return. *Return* is the daily return of VNINDEX. *Policy* is a dummy variable that equals 1 if there is an announcement of a governmental policy related to the Covid-19 pandemic on day t and zero otherwise. *Phase* is a series of dummy variables that represent the different phases of the pandemic in Vietnam. *Special cases* is a dummy variable that equals 1 is there is an announcement of a special infected case or death on day t and zero otherwise. *Cases* and *Deaths* represent the total number of infected cases and death as of day t, respectively. ***, **, and * represent significance levels of 1%, 5%, and 10%, respectively, for the two-tailed T-test

regression includes the control variables and phase fixed effects. Focusing on the second regression, after controlling for any special infected cases announcements, the number of infected cases, the number of deaths, and phase fixed effects, we find that government policy exerts a negative impact on stock market return, although this impact is only significant at the ten percent level. Specifically, an announcement of a government policy results in a 0.45% decrease in daily market return.

Among the control variables, it is worth noting that the announcement of a special infected case results in a decrease of 1.33% in market return, significant at the five percent level. Thus, the impact of a special case announcement is around three times as much as the impact of a government policy announcement.

Next, in our regression model, we replace Policy with a series of dummy variables representing seven types of policies (P_1 to P_7). The regression results, reported in Table 5, are consistent with our univariate results in the sense that we also document the significantly negative impact of P_4 (Border and entry control) and P_7 (Cross-country travel control) on the stock market return. Specifically, as in

Table 5 Regression results for the impact of government policies by types on the VNINDEX

Variables	(1) Return (std.)	(2) Return (std.)
P_1	−0.4674	−0.4341
	(0.3875)	(0.3982)
P_2	−0.2318	−0.2399
	(0.3373)	(0.3642)
P_3	(0.3373)	0.7297
	(1.8640)	(1.9112)
P_4	−2.0071***	−1.9084***
	(0.2415)	(0.2692)
P_5	0.3000	0.3061
	(0.3355)	(0.3490)
P_6	−0.4864	−0.4473
	(0.3965)	(0.4172)
P_7	−2.1410***	−2.1940***
	(0.4383)	(0.4425)
Special cases	−1.5870**	−1.4782**
	(0.6629)	(0.6899)
Cases	0.0008*	0.0016*
	(0.0005)	(0.0009)
Death	−0.0182	−0.0181
	(0.0157)	(0.0210)
AR(1)	−0.2759	−0.2836
	(2.0558)	(2.5290)
MA(1)	0.2386	0.2536
	(2.0797)	(2.5559)
Sigma	1.3341***	1.3302***
	(0.0480)	(0.0481)
Constant	−0.1045	−1.0901
	(0.2116)	(1.1031)
Observations	237	237
Phase fixed effects	No	Yes

This table provides regression results for the relationship between different types of government policies and VNINDEX return. *Return* is the daily return of VNINDEX. P_1 through P_7 is dummy variables that represent different types of government policies related to the Covid-19 pandemic on day t, the coding of policy types can be found in Sect. 5. *Phase* is a series of dummy variables that represent the different phases of the pandemic in Vietnam. *Special cases* is a dummy variable that equals 1 is there is an announcement of a special infected case or death on day t and zero otherwise. *Cases* and *Deaths* represent the total number of infected cases and death as of day t, respectively. ***, **, and * represent significance levels of 1%, 5%, and 10%, respectively, for the two-tailed T-test

Regression (2), the border and entry control policies, on average, cause a decline of 1.91% in market return, whilst the cross-country travel control policies result in a fall of 2.19% in market return. We, however, document no significant impact of P_1 (Outbreak announcements and emergency measures), whereas, in our univariate tests, P_1 marginally affects the VNINDEX return.

7 Conclusion

This chapter examines how government intervention policies affect the stock market as a proxy for economic performance. By analyzing the data of Vietnam, we find that, on average, an announcement of a new policy to control the Covid-19 pandemic has a marginally negative impact on stock market return. However, two types of policies, including border and entry control and cross-country travel control, significantly impact the VNINDEX daily return. Specifically, on average, the border and entry control policies cause a decline of 1.91% in market return, whilst the cross-country travel control policies result in a fall of 2.19%.

We contribute to the limited literature on the economic impact of the Covid-19 pandemic by showing the relationship between pandemic combat and economic development. Specifically, we find that government policies can be very efficient in controlling the Covid-19 pandemic, however, some of them almost cause no damage to the country's economy. We, thus, answer the question that a large number of policy makers are now wondering by showing that there is not always a trade-off between saving lives and economic development.

Our study also provides valuable guidance for other countries in terms of policy decisions during the current fight against the Covid-19 pandemic as well as in future community health crises. While recent literature related to the Covid-19 pandemic mostly shares the focus on the impact of government intervention on controlling the pandemic, limited attention has been paid to investigating the economic impact of such policies on the national economy. We show evidence that different intervention policies can affect the economy differently and suggest that policy makers should not hesitate in adopting rigorous policies to control the pandemic.

References

1. Rowthorn, R., Maciejowski, J.: A cost-benefit analysis of the COVID-19 disease. Oxf. Rev. Econ. Policy 36(S1), S38–S55 (2020)
2. Sebastiani, G.M., Massa, M., Riboli, E.: Covid-19 epidemic in Italy: evolution, projections and impact of government measures. Eur. J. Epidemiol. 35(4), 341–345 (2020)
3. Beaudry, P., Portier, F.: Stock prices, news, and economic fluctuations. Am. Econ. Rev. 96(4), 1293–1307 (2006)
4. Hamilton, J.D., Lin, G.: Stock market volatility and the business cycle. J. Appl. Economet. 11(5), 573–593 (1996)

5. V.M. of Health: Timeline of the COVID-19 pandemic in vietnam. https://ncov.moh.gov.vn/web/guest/dong-thoi-gian (2020). Last accessed on 24 Jan 2021
6. Duong, D.M., Le, V.T., Ha, B.T.T.: Controlling the COVID-19 pandemic in Vietnam: lessons from a limited resource country. Asia Pac. J. Public Health 32(4), 161–162 (2020)
7. Do, T.H., Le, T.H.: Nganh du lich vietnam trong mua dich covid-19 va van de dat ra. http://tapchitaichinh.vn/tai-chinh-kinh-doanh/nganh-du-lich-viet-nam-trong-mua-dich-covid19-va-van-de-dat-ra-329127.html (2020). Last accessed on 17 Jan 2021
8. Shira, D., Associates, Assessing vietnam's economic prospects for foreign investors after COVID-19. https://www.vietnam-briefing.com/news/assessing-vietnams-economic-prospects-foreign-investors-after-covid-19.html (2020). Last accessed on 24 Jan 2021
9. Bank, A.D.: Viet nam's economy remains resilient despite COVID-19 challenges. https://www.adb.org/news/viet-nam-s-economy-remains-resilient-despite-covid-19-challenges (2020). Last accessed on 24 Jan 2021
10. Tooze, A.: Is the Coronavirus crash worse than the 2008 financial crisis?. https://foreignpolicy.com/2020/03/18/coronavirus-economic-crash-2008-financial-crisis-worse (2020). Last accessed on 31 Jan 2021
11. Ding, C.G., Wu, C.H., Chang, P.L.: The influence of government intervention on the trajectory of bank performance during the global financial crisis: a comparative study among Asian economies. J. Financ. Stab. 9(4), 556–564 (2013)
12. La, V.P., Pham, T.H., Ho, M.T., Nguyen, M.H., Nguyen, K.L.P., Vuong, T.T., Nguyen, H.K.T., Tran, T., Khuc, Q., Ho, M.T., Vuong, Q.H.: Policy response, social media and science journalism for the sustainability of the public health system amid the covid-19 outbreak: the vietnam lessons. Sustainability 12(7),(2020)
13. Alexa: Top sites in vietnam. https://www.alexa.com/topsites/countries/VN (2021). Last accessed on 20 Feb 2021
14. Kouzis-Loukas, D.: Learning Scrapy. Packt Publishing Ltd (2016)
15. Singh, B., Dhall, R., Narang, S., Rawat, S.: The outbreak of COVID-19 and stock market responses: an event study and panel data analysis for G-20 countries. Glob. Bus. Rev. (2020)
16. Bouri, E.: Oil volatility shocks and the stock markets of oil-importing MENA economies: a tale from the financial crisis. Energy Econ. 51, 590–598 (2020)
17. V.C. for Disease Control: Covid-19 pandemic situation in vietnam. https://ncov.vncdc.gov.vn/pages/viet-nam-1.html (2020). Last accessed on 31 Jan 2021

Probing Risk of Default in the Market-Leading Islamic Banking Industry: During the Covid-19 Pandemic

Mehreen Mehreen, Maran Marimuthu, Samsul Ariffin Abdul Karim⊙, and Amin Jan

Abstract This study aims to highlight the bankruptcy-related challenges and the reasons behind bank failures for the global Islamic banking industry, and to understand the expected impact of Covid-19 on the financial health of the Islamic banking industry. Moreover, the study shows that going bankrupt is a costly process that affects all stakeholders. Evidence shows that in the case of Islamic banking, a Shariah-based bankruptcy prediction model for apprehending the true bankruptcy prediction is over-sighted. It may further sour the existing uncertain situation for the Islamic banking industry with the additional unfavorable impact of Covid-19. As the subjected pandemic has badly affected the flow of the financial system worldwide, Islamic banking is not isolated. Therefore, efficient bankruptcy prediction models equipped with the latest micro and macro-economic factors are compulsory for the sustainability of the Islamic banking industry.

Keywords Covid-19 · Islamic banking · Bankruptcy prediction · Banks failure · Financial performance

M. Mehreen (✉) · M. Marimuthu
Department of Management and Humanities, Universiti Teknologi PETRONAS, Bandar Seri Iskandar, 32610 Seri Iskandar, Perak Darul Ridzuan, Malaysia
e-mail: mehreen_18001045@utp.edu.my

M. Marimuthu
e-mail: maran.marimuthu@utp.edu.my

S. A. Abdul Karim
Fundamental and Applied Sciences Department and Centre for Systems Engineering (CSE), Institute of Autonomous System, Universiti Teknologi PETRONAS, Bandar Seri Iskandar, 32610 Seri Iskandar, Perak Darul Ridzuan, Malaysia
e-mail: samsul.ariffin@utp.edu.my

A. Jan
Faculty of Hospitality, Tourism, and Wellness, Universiti Malaysia Kelantan, City Campus, 16100 Pengkalan Chepa, Malaysia
e-mail: amin_jan_khan@yahoo.com

© Institute of Technology PETRONAS Sdn Bhd 2022
S. A. Abdul Karim (eds.), *Shifting Economic, Financial and Banking Paradigm*,
Studies in Systems, Decision and Control 382,
https://doi.org/10.1007/978-3-030-79610-5_4

1 Introduction

The financial system is considered the backbone of economies. It offers an organized intermediary facility for funds circulation among various stakeholders (borrowers, lenders, and investors). Financial institutions play an important role in the economic growth and development of economies [6, 13]. For sustainable economic growth and the development of economies, the growth of the banking industry is vital. It is because the banking industry holds a central position in the economies. Sustainable banking industries ensure sustainable economic growth and subsequently reduces the chances of bankruptcy for economies. The global banking industry is mainly composed of conventional and Islamic banking industries.

The Islamic banking system is still quite a new concept, though it is slowly emerging on the financial horizon of the world's financial system [8, 10, 11, 13–17]. Still, based on its low market share in the world's banking assets it requires extraordinary efforts for sustainment [19]. In consonant with that, the key performance indicators affecting the financial health of Islamic banks need to be identified and tackled for providing financial stability and in turn to avoid bankruptcy. For that to happen, dynamic bankruptcy forecasting models are required that must understand the business pulse and must safeguard Islamic banking from the evolving bankruptcy risks.

Ad odds, the currently used bankruptcy forecasting methods, for instance, that of [1, 4] are outdated for multiples reasons. These models consist of the components which support only the financial part of companies [3]. Those models are lagging in considering the glooming bankruptcy risks likewise social and governance-related risks [10]. Furthermore, most of the earlier models were originally developed for manufacturing firms but are still used with little modification on services firms like banks. The same is the agony with Islamic banking, as those outdated models are continuously being used in Islamic banking for bankruptcy prediction [9, 11], which is fundamentally wrong. It is because the Islamic banking system came into existence based on its distinctive nature from its conventional counterparts. Therefore, based on its distinctive nature it should be gauged with dedicated bankruptcy predicting models that follow Shariah principles and Islamic concepts. Otherwise, the process of bankruptcy forecasting would be inaccurate and biased. Apart from the industry need (Islamic banking) for developing dynamic bankruptcy predicting models is also the requirement of the international rating agencies as well. As the ratio analysis, industry health, management plans, strategic decisions, and qualitative factors are strongly considered by the different rating agencies while determining ratings of the companies.

Nowadays, the mixed bankruptcy evaluating models comprising financial and non-financial business attributes likewise considering governance and sustainability are prioritized by consultants, financial bodies, and rating bodies. It is because following the COVID-19 pandemic the dynamics of businesses (including Islamic banks) got changed. The COVID-19 started as a health crisis, swiftly turned into an economic crisis, and now evolving into a humanitarian crisis [18]. Hence, any businesses (including Islamic banks) must now account for those social elements in their governance framework. Ignoring those stakeholders in this current turmoil may negatively affect the firm performance of the Islamic banks, which as a result will expose them towards more bankruptcy. Hence, for Islamic banks to stay competitive in the market they must cope with the evolving business risks and must offer business solutions to the upcoming problems as well. These facts motivate this study to propose a dynamic bankruptcy forecasting model for Islamic banking in compliance with Shariah principles and the latest business trends.

2 Defining "Bankruptcy Forecasting"

Bankruptcy forecasting refers to the process of evaluating various financial distress measures to determine the actual financial health of a bank. Banks use different techniques to predict bankruptcy. [20] classified these techniques broadly into two categories, (a) statistical techniques and (b) intelligent techniques. The first category includes univariate analysis, Multiple Discriminant Analysis (MDA), factor analysis, and Logit regression. The second category consists of neural networks, self-organizing maps, decision trees, operational research techniques such as Data Envelopment Analysis (DEA), and linear programming. Literature is augmented with the use and continuous improvement of the above models for bankruptcy prediction of Banks. Beaver [4], pioneered the work on bankruptcy forecasting by developing a univariate bankruptcy forecasting model. Altman [1], advanced the work by developing a multivariate bankruptcy forecasting model grounded on the Multiple Discriminant Analysis (MDA) techniques. Ohlson [21], further extended the work on bankruptcy forecasting models by using logistic regression. Later on, Altman [2], revised his earlier Z-score and named it a ZETA model. Shumway [24], also extended the bankruptcy forecasting model developed by Altman [1] and named it a Hazard model. The above literature alludes that there are many bankruptcy models available for conventional banks. However, studies related to bankruptcy forecasting in the Islamic banks are found scant in the literature [7, 8, 10–12].

3 Why Do Banks Go Bankrupt?

Financial distress occurs when a bank is facing difficulty to pay the obligations to its creditors. In other words, a bank can go bankrupt, when it fails to return deposits to the customers upon their first demand. Financial distress can lead to bankruptcy if not predicted accurately. The true prediction could mean that the firm is facing only a critical and risky situation. Once predicted, bankruptcy can be avoided by taking the required strategic and operational steps on time. Researchers broadly used the terms default, insolvency, and failure interchangeably. The term insolvency indicates a negative performance when its total liabilities exceed the total assets. The term default is linked with the corporate relationships of stakeholders like creditors, and debtors. In other words, when the bank is unable to pay the obligations to the creditors, the bank can be declared as a defaulter by the regulatory authority. The term failure refers to a situation when the bank is unable to generate optimal returns against the investments.

There are several reasons which can lead banks towards bankruptcy. Many economists claim that financial disasters are the consequences of some distortions, while many stakeholders think that these are the causes instead of consequences. The ultimate goal of an economy is to improve overall economic growth by supporting certain banks and by not supporting inefficient banks. Hence the less productive banks leave the market and are replaced by efficient banks. This process is referred to as 'creative destruction' [23]. Several empirical studies notified that financial distortions and failures are because of some macroeconomic and financial imperfections, for instance, economic fluctuations, low-interest rates, fluctuating inflation rates, and business life cycles [5].

4   What Is the Cost of Going Bankrupt?

Going bankrupt is a costly process and it excessively affects the stakeholders. The effect is shown more in the value of stocks. In line with this effect, the overall value of the bank goes down [22]. Literature suggests that there are two types of bankruptcy costs, direct and indirect costs [25]. Some of these costs likewise the cost of risk and mental stress of the stakeholders can be predicted but is not possible to quantify. These are considered to be the parts of non-quantifiable costs. Moreover, researchers have attempted to measure the cost of bankruptcy and the traditional theory of corporate finance helped to make a precise distinction between the direct and indirect costs of financial distress. Administrative formalities like administrative fees, lawyers' and accountants' fees are considered as direct costs of bankruptcy. The indirect cost of bankruptcy includes profits loss, lost sales, inability to issue securities, psychological stress, and mental illness of the affected stakeholders. All these indirect costs are unquantifiable. Despite the fact, there may be other reasons for the costly threat of financial distress. The firm's history in terms of its health and financial behavior can affect the decisions of customers and suppliers towards the firm. The above reasons can influence suppliers and the existing banking staff towards finding new and more secure jobs. And therefore, it will further affect the financial position of banks in a negative direction. These facts allude to that the cost of bankruptcy can be very detrimental in terms of physical capital and social capital.

5   Global Islamic Banking Industry

The term Islamic banking refers to a system that is based on Islamic principles and guided by Islamic economics. The fundamental principles of Islamic banking are the sharing of profit and losses, and the prohibition of interest during collection and payment by the lenders and investors. According to the Islamic Finance Development report 2020 pg. 22, the total global Islamic finance assets accounts for USD 2.88 trillion. Islamic banking is the main component of Islamic finance with a share of 1993 billion USD, followed by Sukuk with 538 USD billion, other Islamic financial institutions with 140 billion USD, and Takaful retains 51 Billion USD share out of the total 2.88 Trillion USD (IFDI 2020, p. 8)[1]. The breakdown of global Islamic finance is depicted in Fig. 1.

According to the Islamic Finance Development report 2020, currently, there is a total of 526 Islamic banks operating in the world, its assets worth 2 trillion USD and it accounts for 6% of the global banking industry (IFDI 2020, p. 26)[3]. The report extended that the Islamic banking sector holds 69 percent assets share (almost 2 trillion) in the Islamic financial industry. Islamic banks lead the market share in the major Muslim countries [11]. The following Fig. 2 shows the percentage of various Muslim countries among the total Islamic banking assets.

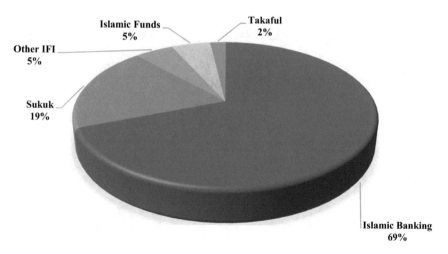

Fig. 1 Breakdown of global islamic finance (IFDI 2020)[2]

---

[1] https://icd-ps.org/uploads/files/ICD-Refinitiv%20IFDI%20Report%2020201607502893_2100. pdf.

[2] https://icd-ps.org/uploads/files/ICD-Refinitiv%20IFDI%20Report%2020201607502893_2100. pdf.

[3] https://icd-ps.org/uploads/files/ICD-Refinitiv%20IFDI%20Report%2020201607502893_2100. pdf.

6   Market Leading Islamic Banking Countries

The following Fig. 2 shows the breakdown of Islamic Banking Islamic assets in terms of an individual country's share.

Figure 2 shows the share of the market-leading Islamic banking counties ranked by global banking assets. It shows that the top tier Islamic banking countries are Iran which holds 29 percent of the global Islamic banking assets followed by Saudi Arabia with 25 percent, Malaysia 11 percent, UAE 9 percent, and Kuwait 6 percent each. The above top five countries hold 80 percent of the global Islamic banking assets. The significant share signifies that the sustainability of the Islamic banking industry from these countries is highly required. In case of any financial turmoil among the Islamic banking industry of these countries will, in turn, affect the global Islamic banking industry. Therefore, it is vital to evaluate its bankruptcy profile with the latest bankruptcy forecasting tools/methods/framework/indexes. The agony is that the bankruptcy forecasting models based on the Shariah principle are over sighted and received limited attention. At further odds, the existing conventional bankruptcy forecasting models are outdated for many reasons. Therefore, considering this nicety, multi-dimensional bankruptcy forecasting models are required for evaluating the true bankruptcy profile of the market-leading Islamic banking industry.

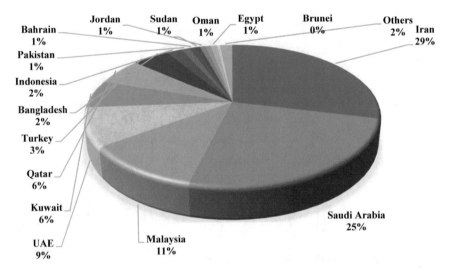

Fig. 2   Individual country share in the global islamic banking assets [4]

---

[4] Islamic Financial Services Industry Stability Report 2020, p. 15.

7 Impact of Covid-19 on the Islamic Banking in Terms of Bankruptcy Forecasting

The COVID-19 crisis presents Islamic banks with a unique opportunity to turn and strengthen their long-term positions. The driving factors include Islamic corporate governance, socially responsible investment, the Islamic finance industry's social position, and a recommitment to sustainability, the crisis is changing the dynamics of the industry and putting new prospects for the Islamic banking industry in motion. The pandemic has demonstrated the importance of stress-testing s a risk management strategy. An Islamic bank's overall governance should include stress testing. Solvency and liquidity stress testing can include both extreme and realistic scenarios, as well as reverse stress testing (the stresses that would cause a bank to fail). The pandemic insights the Islamic banks for upgrading their traditional bankruptcy forecasting model through integrating the latest business trends as explained below.

7.1 Compliance with Islamic Corporate Governance Measures for Mitigating Bankruptcy

The pandemic urged Islamic banks to update their overall performance model (including bankruptcy forecasting models). The previous bankruptcy forecasting models in the Islamic banks were based only on the financial ratios. The time demands to integrate industry-specific (Islamic corporate governance) measures to the bankruptcy evaluation. It is because, Shariah committee exists in every Islamic bank, but the bankruptcy knowledge of Shariah scholars is minimal. Shariah Committee members must be given due training to illuminate them about if any Shariah ruling may trigger bankruptcy? Therefore, integrating bankruptcy surveillance into the Islamic corporate governance framework will reduce the chances of bankruptcy during this time of crisis. And hence, the following proposition is proposed.

P_1: *Integrating bankruptcy surveillance into the Islamic corporate governance framework will reduce the chances of bankruptcy for the Islamic banks during the pandemic.*

7.2 Compliance with Corporate Sustainability Measures for Mitigating Bankruptcy

The pandemic has affected multiple stakeholders across various business sectors (including Islamic banks). Specifically, the Triple Bottom Line TBL (People, profit, and the planet) got severely affected by the pandemic [18]. Against that background, those Islamic banks that addressees the triple bottom line will mitigate the distressing impact of a pandemic on the Islamic banking stakeholders. Addressing deprived stakeholders at this point of crisis will earn Islamic banks tangible and intangible benefits. For instance, and a direct impact, supporting skilled micro-entrepreneurs who lost their jobs during the pandemic will regenerate economic activities in the economy, it will create more cash flow, and the increased cash flow will increase the demands for banking services. Hence, the bank will receive higher interest on the provided loans. And at the same time will increase the deposits of banks. As an indirect impact, the commitment of Islamic banks will improve their philanthropic image, which in turn will improve their sustainability ratings as well. And the higher sustainability ratings allow the Islamic banks to apply for a higher loan. Better funds will improve the financial performance of Islamic banks. Subsequently, better firm performance will allow Islamic banks to be less bankrupt. In the same vein, this study proposes the following proposition.

P_2: *Compliance with sustainable practices will mitigate bankruptcy chances for the Islamic banks during the pandemic.*

8 Compulsion of Developing a Multi-dimensional Bankruptcy Forecasting Model for the Market-Leading Islamic Banking Countries Following the Covid-19 Pandemic

The legitimacy of developing a multi-dimensional bankruptcy forecasting model for the market-leading Islamic banking countries is backed by multiple reasons. Firstly, as shown in the above Fig. 4.2 the top-five market-leading Islamic banking countries account for 80 percent of the global Islamic banking assets. In case of any turmoil among the Islamic banking industry of these countries will, in turn, affect the global Islamic banking industry. Secondly, outdated models are continuously being used in Islamic banking for bankruptcy prediction [9, 11], which is fundamentally wrong. It is because the Islamic banking system came into existence based on its distinctive nature from conventional banks. Therefore, it should be gauged with dedicated bankruptcy predicting models that follow Shariah principles and Islamic concepts. Otherwise, the process of bankruptcy forecasting would be inaccurate and biased. Thirdly, the Covid-19 pandemic changed prior business methods for performance evaluation. Now every business sector (including Islamic banks) is required to update their business models (including bankruptcy forecasting model) by integrating more business aspects apart from financial aspects only. In the current situation, the mixed bankruptcy evaluating models comprising financial and non-financial business attributes likewise considering governance and sustainability are prioritized by consultants, financial bodies, and rating agencies. Hence, in the short (a) the huge market share, (b) outdated bankruptcy forecasting models, and (c) the urgency for upgrading bankruptcy forecasting following the Covid-19 legitimizes this study for the development of a multi-dimensional bankruptcy forecasting model in line with Shariah principles.

Most of the prior bankruptcy forecasting models for instance [1, 4], were based on financial ratios and it lagged in integrating vital business elements such as governance and sustainability. Considering this outdating factor and the subsequent urge for updating forecasting methods following the pandemic this study proposes the role of (a) Islamic corporate governance and (b) corporate sustainability to be part of the deemed multi-dimensional bankruptcy forecasting model. Considering these factors will provide Shariah backing in the process of bankruptcy forecasting. At the same time, it will pacify multiple stakeholders such as (economic, environmental, and social) who got affected by the pandemic in various ways.

The agency theory supports the role of Islamic corporate governance variables in predicting bankruptcy. The theory deals with the principle-agent relationship. The shareholders are the principle, and the management are considered as the agent. The agents are supposed to work for the maximization of the shareholder's value. If the interest of the shareholders is aligned with the shareholder's value maximization, holistically it will increase the value of the firm as well. And in a way, better firm performance will reduce the chances of bankruptcy and vice versa. Similarly, the stakeholders' theory supports the role of using sustainability items in bankruptcy prediction. The stakeholders' theory argues that the firm value will be maximized when it addresses multiple stakeholders' such as economic, social, environmental. And once the firm value is improved, it reduces the chances of bankruptcy and vice versa.

9 Summary

This study highlights bankruptcy-related challenges and the reasons behind banks' failures for the global Islamic banking industry and understands the expected impact of Covid-19 on the financial health of the Islamic banking industry. In the same vein, this study defined and explained bankruptcy forecasting, reasons for why banks go bankrupt, and its associated detrimental cots. In consonant to the main theme, this study explained the global Islamic banking profile with a share of an individual country as well. The later part of this study illuminates the possible impact of the Covid-19 pandemic on the Islamic banks and elaborate on the required compliance with sustainability practices and Islamic corporate governance for mitigating bankruptcy chances. In the same line, this study proposed two propositions. At last, this study explains why there is a need to propose a multi-dimensional bankruptcy forecasting model for the market-leading Islamic banking countries. This study anticipates that these transformations will pacify multiple stakeholders towards better firm performance. And the better firm performance will mitigate the chances of bankruptcy for the Islamic banks during the pandemic.

References

1. Altman, E.I.: Financial ratios, discriminant analysis and the prediction of corporate bankruptcy. J. Financ. **23**, 589–609 (1968)
2. Altman, E.I.: Predicting financial distress of companies: revisiting the Z-score and ZETA® models. In: Handbook of research methods and applications in empirical finance. Edward Elgar Publishing (2013)
3. Altman, E.I.: A fifty-year retrospective on credit risk models, the Altman Z-score family of models and their applications to financial markets and managerial strategies. J. Credit Risk **14** (2018)
4. Beaver, W.H.: Financial ratios as predictors of failure. J. Account. Res. 71–111 (1966)
5. Bhattacharjee, A., Higson, C., Holly, S., Kattuman, P.: Macroeconomic instability and business exit: determinants of failures and acquisitions of UK firms. Economica **76**, 108–131 (2009)
6. Hanif, M., Tariq, M., Tahir, A., Momeneen, W.U.: Comparative performance study of conventional and islamic banking in Pakistan. Int. Res. J. Financ. Econ. (2012)
7. Husna, H.N., Rahman, R.A.: Financial distress-Detection model for Islamic banks. Int. J. Trade Econ. Financ. **3**, 158 (2012)
8. Jan, A., Marimuthu, M.: Altman model and bankruptcy profile of Islamic banking industry a comparative analysis on financial performance (2015a)
9. Jan, A., Marimuthu, M.: Altman model and bankruptcy profile of Islamic banking industry: a comparative analysis on financial performance (2015b)
10. Jan, A., Marimuthu, M.: Bankruptcy and sustainability: a conceptual review on islamic banking industry (2015c)
11. Jan, A., Marimuthu, M.: Sustainability profile of islamic banking industry: Evidence from world top five islamic banking countries (2015d)
12. Jan, A., Marimuthu, M.: Bankruptcy profile of foreign versus domestic islamic banks of Malaysia: a post crisis period analysis (2016a)
13. Jan, A., Marimuthu, M.: Bankruptcy profile of foreign vs. domestic islamic banks of Malaysia: a post crisis period analysis. Int. J. Econ. Financ. Issues **6** (2016b)
14. Jan, A., Marimuthu, M., Bin Mohd, M.P., Isa, M.: The nexus of sustainability practices and financial performance: from the perspective of islamic banking. J. Clean. Prod. **228**, 703–717 (2019a)
15. Jan, A., Marimuthu, M., Bin Mohd, M.P., Isa, M., Shad, M.K.: Bankruptcy forecasting and economic sustainability profile of the market leading islamic banking countries. Int. J. Asian Bus. Inform. Manag. (IJABIM) **10**, 73–90 (2019b)
16. Jan, A., Marimuthu, M., Hassan, R.: Sustainable business practices and firm's financial performance in islamic banking: under the moderating role of islamic corporate governance. Sustainability **11**, 6606 (2019)
17. Jan, A., Marimuthu, M., Pisol, M., Isa, M., Albinsson, P.: Sustainability practices and banks financial performance: a conceptual review from the islamic banking industry in Malaysia. Int. J. Bus. Manag. **13** (2018)
18. Jan, A., Mata, M.N., Albinsson, P.A., Martins, J.M., Hassan, R.B., Mata, P.N.: Alignment of islamic banking Sustainability Indicators with sustainable development goals: policy recommendations for Addressing the COVID-19 pandemic. Sustainability **13**, 2607 (2021)
19. Khan, I., Khan, M., Tahir, M.: Performance comparison of Islamic and conventional banks: empirical evidence from Pakistan. Int. J. Islamic Middle East. Financ. Manag. (2017)
20. Kumar, P.R., Ravi, V.: Bankruptcy prediction in banks and firms via statistical and intelligent techniques–A review. Eur. J. Oper. Res. **180**, 1–28 (2007)
21. Ohlson, J.A.: Financial ratios and the probabilistic prediction of bankruptcy. J. Account. Res. 109–131 (1980)
22. Patti, A.: Financial distress and bankruptcy: tools to preserve the soundness of the financial system (2015)
23. Reistadaasen, M.: Applying Altman's Z-score to the financial crisis: an empirical study of (2011)

24. Shumway, T.: Forecasting bankruptcy more accurately: a simple hazard model. J. Bus. **74**, 101–124 (2001)
25. Warner, J.B.: Bankruptcy costs: some evidence. J. Financ. **32**, 337–347 (1977)

Covid-19 and Cryptocurrency Markets Integration

Bakri Abdul Karim, Aisyah Abdul Rahman, Syajarul Imna Mohd Amin, and Norlin Khalid

Abstract This paper examines the impact of COVID-19 on the integration and dynamic linkages of the cryptocurrencies (Bitcoin, Ethereum, Litecoin, XRP and Stellar). ARDL bound test approach and Granger causality tests are used in this study for the period from 17 April 2019 to 15 September 2020. We found evidence of no cointegration among the cryptocurrencies in both pre- and during COVID-19. Thus, the cryptocurrencies market offers an ample opportunity for the potential benefits from portfolio diversification and hedging strategies even during the COVID-19 pandemic. In addition, Granger causality tests show that Bitcoin is the most influential cryptocurrency in the short-run. The findings of this study may have implications for crypto-investors, international investors and fund managers who want to diversify their investments in cryptocurrencies.

Keywords Covid-19, cryptocurrencies · Market integration · Portfolio diversification

1 Introduction

The emergence of COVID-19 in early 2020 has been causing chaos around the globe. Later it was declared as a pandemic by WHO in March 2020. As of 26 November 2020, the number of confirmed cases has been reached to 59,204,902 with 1,397,139 deaths globally (WHO 2020). The pandemic has been causing huge impacts on both social and economic activities. Goodell [13] noted that COVID-19 is causing unprecedented global economic harm and wide-ranging impact on financial sectors. Phan and Narayan [29] argued that COVID-19 is the father of all

B. A. Karim (✉)
Faculty of Economics and Business, Universiti Malaysia Sarawak (Malaysia), 94300 Kota Samarahan, Sarawak, Malaysia
e-mail: akbakri@unimas.my

A. Abdul Rahman · S. I. Mohd Amin · N. Khalid
Faculty of Economics and Management, Universiti Kebangsaan Malaysia (UKM), Bangi, Selangor, Malaysia

© Institute of Technology PETRONAS Sdn Bhd 2022
S. A. Abdul Karim (eds.), *Shifting Economic, Financial and Banking Paradigm*,
Studies in Systems, Decision and Control 382,
https://doi.org/10.1007/978-3-030-79610-5_5

fears that have engulfed global financial and economic systems. Due to the global pandemic of COVID-19, financial and commodity prices around the world have plunged tremendously [16], unemployment reached highest level in many countries [21], sales declined, production reduced and companies were in serious financial burden [23].

Investors are always looking for alternative assets to reduce their equity investments' downside risk particularly during this catastrophic market environment. The emergence of cryptocurrency has attracted a lot of discussion as they have a potential in hedging and investments [4]. The popularity and exponential growth of cryptocurrencies has attracted media attention, researchers and policy makers since the launch of Bitcoin in 2009. The cryptocurrencies market has rapidly become a vital component in the global financial market with the value grew from USD17.7 billion early 2017 to USD700 billion in early 2018 [17]. Investors and fund managers have seen cryptocurrencies as an investable asset with the ability to generate high return regardless of their extreme volatility [17]. Besides a store of values during market turmoil period, cryptocurrencies also can be a source of portfolio diversification [10]. In addition, Liu [24] also noted that the inclusion of cryptocurrency can enhanced portfolio benefits.

Although empirical studies on cryptocurrencies are rapidly increasing, studies on the impacts of COVID-19 on the co-movements and integration amongst cryptocurrencies are rather limited. For example, Yousaf and Ali [32] studied the return and volatility spillover between three cryptocurrencies (Bitcoin, Ethereum and Litecoin) during the pre-COVID-19 and COVID-19 period. While Iqbal et al. [16] examined the impact of COVID-19 outbreak on the top 10 cryptocurrencies' returns and found that majority of them performed better against the pandemic. However, Yarovaya et al. [30] provided evidence that cryptocurrencies were the riskiest in the long-term, with more than a 50% decline in value coupled with high degrees of persistence during COVID-19. In addition, Lahmiri and Bekiros [23] found cryptocurrencies entrench instability and irregularity and therefore investing in this market during big crisis could be considered riskier compared to equities. Against this backdrop, this paper aims to examine the impact of COVID-19 on the cryptocurrency markets integration and their dynamic linkages. It is important to evaluate how the cryptocurrency assets behave and act as a hedging and diversifications instruments in this pandemic times.

This study contributes to the existing literature on the cryptocurrencies in three ways. First, we evaluate how the major cryptocurrencies' return and correlations respond in the pre- and during COVID-19. Second, we identify the differences in the cryptocurrencies' integration and their dynamic linkages in both normal and pandemic periods. Third, this study relies on ARDL bound test approach proposed by Pesaran et al. [27]. Narayan et al. [26] indicated ARDL model has more advantage as we can state which variable is the dependent variable from the F-test when integration exists. In addition, Pesaran and Shin [28] and Narayan and Smyth [25] show that with the ARDL framework, the ordinary least squares estimators of are super-consistent in small sample sizes. The findings of this study may have implications for crypto-investors, international investors and fund managers who want to diversify their investments in cryptocurrencies.

The rest of the paper is organized as follows. Section 2 provides some related literature review. Section 3 describes empirical framework and data description. Section 4 reports empirical results and discussion. Lastly, Sect. 5 presents some concluding remarks.

2 Literature Review

Corbet et al. [8] have provided a good systematic literature review on major topics of cryptocurrencies such as pricing bubble, regulation, cybercrime, diversification and efficiency. In terms of co-movement of cryptocurrencies, few studies are worth to discuss. Beneki et al. [2] examined the relationship between Bitcoin and Ethereum. They found significant in the time-varying correlation and a positive response of Bitcoin volatility on a positive volatility shock on Ethereum returns. Using DCCA approach, Ferreira and Pereira [12] attempted to evaluate the contagion effect amongst Bitcoin and other major cryptocurrencies and they concluded that the cryp-tocurrencies markets are more integrated. In addition, Ji et al. [17] investigated the linkages of returns and volatility across six large cryptocurrencies for the period between August 2015 to February 2018. The results show that returns shocks from Litecoin and Bitcoin has the most effect on other cryptocurrencies while Ethereum and Dash show very weak integration. Yi et al. [31] analyze the volatility connected-ness between the 52 cryptocurrencies. They found that a volatility transmission from Bitcoin to other cryptocurrencies and several other cryptocurrencies also transmit strong volatility effects. Moreover, both Katsiampa et al. [20] and Canh et al. [5] also found a significant volatility transmission amongst cryptocurrencies.

Another study, Corbet et al. [9] examined the dynamic relationships between three cryptocurrencies and several financial assets. They found evidence that cryptocurren-cies could provide diversification benefits for investors particularly with short term investment horizons. Bouri et al. [3] examined the dynamic conditional correlation between Bitcoin and four major world stock indices, bond, oil, gold, commodity index and the US dollar. They found that Bitcoin is able to provide a good diver-sification. Moreover, Baur et al. [1] show that Bitcoin displayed different return, volatility and correlation features compared to other assets and it possessed the same hedging abilities as gold. However Conlon et al. [7] found that Bitcoin and Ethereum are not a safe haven for the majority of international equity markets. In the context of COVID-19 and cryptocurrencies, Corbet et al. [10] showed that the volatility connection between the main Chinese stock markets and Bitcoin evolved signifi-cantly during the COVID-19 period. While Iqbal et al. [16] examined the impacts of COVID-19 outbreak on the top 10 cryptocurrencies returns. They found that majority of the cryptocurrencies performed better against small shocks of COVID-19. In addi-tion, they also found that Bitcoin, ADA, CRO and Ethereum performed better against the negative effects of the highest intensity of COVID-19. Using VAR-AGARCH model, Yousaf and Ali [32] found return spillover vary during the pre-COVID-19 and COVID-19 period for cryptocurrencies. In a recent study, Lahmiri and Bekiros [23] provided evidence that during the COVID-19 pandemic period, the level of stability in cryptocurrency markets has significantly diminished.

3 Empirical Framework and Data Description

3.1 Empirical Framewok

The study employs the ARDL bounds test proposed by Pesaran et al. [27] to investigate the integration between the cryptocurrencies. Unlike other techniques such as the Johansen and Juselius [18], the ARDL approach does not require the pre-testing of the variables included in the model. Another benefit of ARDL is it takes a sufficient number of lags in a general-to-specific modelling to capture the data-generating process [19]. The ARDL procedure involves two phases. In the first phase, we establish a long-run relationship exists among the cryptocurrencies. The second phase, if the variables are cointegrated, then we estimate the long- and short-run coefficients of equations. In Pesaran et al. [27], details of the mathematical derivation of the long- and short-run parameters can be found. For cointegration analysis, Δyt is modelled as a conditional Error-Correction Model as follows (assuming two variables):

$$\Delta y_t = \alpha_0 + \lambda_y y_{t-1} + \pi_x x_{t-1} + \sum_{i=1}^{p} \theta_i \Delta y_{t-i} + \sum_{j=0}^{p} \phi_j \Delta x_{t-i} + \mu_t \qquad (1)$$

where, λ y and πx are long-run multipliers. Lagged values of Δyt and current and lagged values of Δxt are employed to model the short-run dynamics. The existence of integration is drawn via restricting all estimated coefficients of lagged level variables equal to zero (the null hypothesis H0 $:= \lambda y = \pi x = 0$ against the alternative, hypothesis Ha : $\lambda y \neq \pi x \neq 0$). To test these hypotheses, we used the critical values bounds as tabulated in Pesaran et al. [27] under case III with unrestricted intercepts and no trend and number of regressors, k are 4. If the computed F-statistic is below than lower bound critical value, the null hypothesis of no integration is not rejected. However, if the computed F-statistics is more than upper bound critical value, the null hypothesis is rejected and thus there exists a cointegration relationship among variables. Nevertheless, the result is inconclusive if the computed F-value falls within lower and upper bound critical values.

In addition, we also run Granger [14] causality test to examine the dynamic linkages among the cryptocurrencies. For a cointegrated among time series, there must be causality among them at least in one direction [15]. Moreover, for any cointegrated variables, error correction term must be included in the model. This model is identified as a vector error correction model (VECM). Engle and Granger [11] showed that omitting this error correction term in the model shall lead to model mis-specification. However, if the variables are not cointegrated, it is appropriate to use VAR model in first differenced.

3.2 Data Description

This study uses daily data of the cryptocurrencies namely Bitcoin (BIT), Ethereum (ETHE), Litecoin (LITE), XRP and Stellar (STEL) covering the period from 17 April 2019 to 15 September 2020. We then divide the data into pre-COVID-19 and during COVID-19 periods. The pre-COVID-19 period covers from 17 April 2019 to 31 December 2019. While, the COVID-19 period covers from 1 January 2020 to 15 September 2020. The cryptocurrencies data were collected from https://coinmarke tcap.com. All series are transformed into natural logarithm.

4 Empirical Results

Table 1 provides some descriptive statistics of the variables such as sample mean, maximum, minimum, standard deviations, skewness and kurtosis. During pre-COVID-19 only Bitcoin recorded positive average daily returns at 0.12% while other cryptocurrencies recorded negative average daily return. Interestingly, during COVID-19 all cryptocurrencies recorded positive average daily return where Ethereum was the highest at 0.4%. Bitcoin is considered very stable with positive return in both periods. Kristoufek [22] noted that the cryptocurrency such as Bitcoin was independent from the global financial system as its prices were determined by non-economic and financial factors. We also can see that the standard deviation of all cryptocurrencies increased during COVID-19. It means that all cryptocurrencies experienced an increase of returns volatility during COVID-19. Ethereum is found to be more volatile in both periods. All cryptocurrencies have larger excess kurtosis than a normal distribution.

Table 2 shows the results from standard correlation of coefficient. During pre-COVID-19, the results show that Ethereum has low correlations with other cryptocurrencies. The highest correlation is between XRP and Stellar at 0.81. Compared to pre-COVID-19, there is a marked increase in correlation among all pairs during COVID-19. Bitcoin-Litecoin and XRP-Litecoin pairs recorded the highest correlation at 0.89. The significant increase in the correlation coefficient among the cryptocurrencies indicates that there are short-term co-movements among the cryptocurrencies thus suggesting that short-term diversification benefits reduced in these assets during COVID-19.

Table 1 Summary statistics of the cryptocurrencies returns

		BIT	ETHE	LITE	STEL	XRP
Pre-COVID19	Mean	0.0012	−0.0051	−0.0025	−0.0037	−0.0021
	Maximum	0.1448	0.1661	0.1451	0.2654	0.2286
	Minimum	−0.1518	−0.2069	−0.1803	−0.1377	−0.1342
	Std. Dev.	0.0383	0.0526	0.0441	0.0442	0.0386
	Skewness	0.0704	−0.3398	0.0537	1.1432	0.5111
	Kurtosis	6.0507	4.8094	5.1296	9.6356	8.8552
	Jarque-Bera	100.26[***]	40.15[***]	48.87[***]	529.52[***]	379.78[***]
COVID19	Mean	0.0016	0.0040	0.0005	0.0021	0.0009
	Maximum	0.1671	0.2516	0.1910	0.1673	0.1426
	Minimum	−0.4647	−0.5055	−0.4490	−0.4100	−0.3990
	Std. Dev.	0.0438	0.0532	0.0529	0.0509	0.0447
	Skewness	−4.3946	−3.4447	−2.3815	−1.8849	−2.5981
	Kurtosis	52.5449	37.0356	23.8924	19.5212	27.5922
	Jarque-Bera	27218.40[***]	12963.26[***]	4936.16[***]	3086.96[***]	6791.56[***]
Full Sample	Mean	0.0014	−0.0006	−0.0010	−0.0008	−0.0006
	Maximum	0.1671	0.2516	0.1910	0.2654	0.2286
	Minimum	−0.4647	−0.5055	−0.4490	−0.4100	−0.3990
	Std. Dev.	0.0411	0.0530	0.0486	0.0477	0.0417
	Skewness	−2.6169	−1.8933	−1.4895	−0.6616	−1.3743
	Kurtosis	35.9424	20.5111	18.1578	15.7831	21.0756
	Jarque-Bera	23967.17[***]	6914.37[***]	5140.56[***]	3557.79[***]	7200.98[***]

Notes [***] *denotes significant at 1%*

Table 2 Correlation of cryptocurrencies returns

	Pre-COVID19	COVID19	Full Sample
BIT-ETHE	− 0.01	*0.56*	0.29
BIT-LITE	0.73	*0.89*	0.82
BIT-STEL	0.59	*0.78*	0.69
BIT-XRP	0.68	*0.85*	0.78
ETHE-LITE	0.01	*0.52*	0.29
ETHE-STEL	−0.07	*0.52*	0.25
ETHE-XRP	−0.05	*0.54*	0.27
LITE-STEL	0.72	*0.81*	0.77
LITE-XRP	0.76	*0.89*	0.84
STEL-XRP	0.81	*0.85*	0.83

The results of the ARDL for integration are reported in Table 3. Narayan et al. [26] indicated that another advantage of the ARDL model is that we can state which variable is the dependent variable from the F-test when integration exists. Pesaran and Shin [28] also noted that ARDL model requires a priori knowledge of the orders of the extended ARDL that is sufficient to simultaneously correct for residual serial correlation and the problem of endogenous regressors. Thus, the order of the distributed lag on the dependent variable and the regressors is selected using AIC.

For the pre-COVID-19 period, the F-test show that the null hypothesis of no integration among cryptocurrencies cannot be rejected because F-statistics are below the lower bound critical value. The same results are also found during COVID-19 and full-sample period where all F-statistics are below lower bound critical value. The results indicate that the cryptocurrencies are not integrated in the long-run thus provide potential benefits for diversification. COVID-19 seems does not affect the long-run co-movement among the cryptocurrencies.

Next, we proceed to examine dynamic linkages amongst the cryptocurrencies using Granger causality. Since all variables are not integrated in both periods, we use VAR model in first differenced. Table 4 reports the results of Granger causality. During pre-COVID-19, there seems to be short-run unidirectional causalities relationship running from Bitcoin to Litecoin and Stellar, from Stellar to Ethereum, from Litecoin to Stellar and from XRP to Stellar. However, the number of causalities reduced during COVID-19 where unidirectional causality only found from Bitcoin to Stellar and XRP and from Stellar to Ethereum. Consistent with Ji et al. [17], we also found that Bitcoin has the most effect on other cryptocurrencies. However, Yi et al. [31] found that Bitcoin is not the dominant transmitter of volatility to other cryptocurrencies.

The evidence of no cointegration among these markets, indicate that the cryptocurrencies offer potential benefits of portfolio diversification. The results are in line with Baur et al. [1], Corbet et al. [9], Ciaian and Rajcaniova [6] and Bouri et al. [3]. For instance, Baur et al. [1] provided evidence that cryptocurrency such as Bitcoin exhibited different return, volatility and correlation features compared to other assets and it possessed the same hedging abilities as gold. Moreover, Bouri et al. [4] found evidence that many cryptocurrencies were potentially valuable asset class and safe-havens. Using three most popular cryptocurrencies and a variety of other financial assets, Corbet et al. [9] found evidence of the relative isolations of cryptocurrencies form other financial and economic assets. Besides that, Ciaian and Rajcaniova [6] examined the interdependencies amongst Bitcoin and Altcoin markets for the period between 2013 and 2016. They found that both markets are interdependent. Another study, Bouri et al. [3] revealed that Bitcoin does act as a hedge against uncertainty.

5 Conclusion

In this study, we examine the impact of COVID-19 on the integration and dynamic linkages amongst cryptocurrencies (Bitcoin, Ethereum, Ripple, Litecoin and Stellar)

Table 3 ARDL Estimation

Period	Equation	The computed F-Statistics	Outcome
Pre Covid19	F(BIT/ETH, LIT, XRP, STE)	2.38	No integration
	F(ETH/BIT, LIT, XRP, STE)	1.85	No integration
	F(LIT/BIT, ETH, XRP, STE)	1.39	No integration
	F(XRP/BIT, ETH, LIT, STE)	1.64	No integration
	F(STE/BIT, ETH, LIT, RIP)	1.68	No integration
During Covid19	F(BIT/ETH, LIT, XRP, STE)	2.44	No integration
	F(ETH/BIT, LIT, XRP, STE)	1.65	No integration
	F(LIT/BIT, ETH, XRP, STE)	1.76	No integration
	F(XRP/BIT, ETH, LIT, STE)	1.64	No integration
	F(STE/BIT, ETH, LIT, RIP)	1.16	No integration
Full Sample	F(BIT/ETH, LIT, XRP, STE)	2.16	No integration
	F(ETH/BIT, LIT, XRP, STE)	2.06	No integration
	F(LIT/BIT, ETH, XRP, STE)	1.28	No integration
	F(XRP/BIT, ETH, LIT, STE)	1.20	No integration

(continued)

Table 3 (continued)

Period	Equation	The computed F-Statistics	Outcome
	F(STE/BIT, ETH, LIT, XRP)	1.24	No integration

Note The relevant critical value bounds are obtained from Pesaran et al. [27], where the critical values in the case of 4 regressors are 2.86–4.01 at the 95% significance level and 2.45–3.52 at the 90% significance level

relying on ARDL bound test approach and Granger causality tests over the period from 17 April 2019 to 15 September 2020. The results show that there is no cointegration found amongst the cryptocurrencies in both pre-COVID-19 and COVID-19 periods thus indicating that potential benefits of portfolio diversification and hedging strategies in long-run even during COVID-19. COVID-19 also does not influence significantly short-run dynamic linkages amongst the cryptocurrencies. The results also show the influential role of Bitcoin on other cryptocurrencies. Cryptocurrencies extended the variety of investment and risk management strategies available for investors.

As the cryptocurrencies' markets are not tied together in the long-run this indicates that the cryptocurrencies markets are efficient following the Efficient Market Hypothesis (EMH). This implies that each variable does not contains information on the common stochastic trends thus investors cannot explore arbitrage profits utilizing information on other cryptocurrencies prices. Moreover, the findings of this study may have implications for crypto-investors, international investors and fund managers who want to diversify their investments in cryptocurrencies. Therefore, cryptocurrency investors and fund managers can restructure their investment

Table 4 Granger Causality

	Pre-COVID19	COVID19	Full Sample
BIT-ETHE	BIT =/= ETHE	BIT =/= ETHE	BIT =/= ETHE
BIT-LITE	**BIT ==> LITE**	BIT =/= LITE	**BIT ==> LITE**
BIT-STEL	**BIT ==> STEL**	**BIT ==> STEL**	**BIT ==> STEL**
BIT-XRP	BIT =/= XRP	**BIT ==> XRP**	**BIT ==> XRP**
ETHE-LITE	ETHE =/= LITE	ETHE =/= LITE	ETHE =/= LITE
ETHE-STEL	**ETHE <==STEL**	**ETHE <==STEL**	**ETHE <==STEL**
ETHE-XRP	ETHE =/= XRP	ETHE =/= XRP	ETHE =/= XRP
LITE-STEL	**LITE ==> STEL**	LITE =/= STEL	LITE =/= STEL
LITE-XRP	LITE =/= XRP	LITE =/= XRP	LITE =/= XRP
STEL-XRP	**STEL <== XRP**	STEL =/= XRP	**STEL <== XRP**

strategies to maximize the risk-returns trade-off using right combination of cryptocurrencies markets. Moreover, Ji et al. [17] noted that cryptocurrencies are investable asset class capable of producing high returns notwithstanding their extreme volatility.

References

1. Baur, D.G., Dimpfl, T., Kuck, K.: Bitcoin, gold and the US dollar–a replication and extension. Financ. Res. Lett. **25**, 103–110 (2018)
2. Beneki, C., Koulis, A., Kyriazis, N.A., Papadamou, S.: Investigating volatility transmission and hedging properties between Bitcoin and Ethereum. Res. Int. Bus. Financ. **48**, 219–227 (2019)
3. Bouri, E., Molnár, P., Azzi, G., Roubaud, D., Hagfors, L.I.: On the hedge and safe haven properties of Bitcoin: Is it really more than a diversifier? Financ. Res. Lett. **20**, 192–198 (2017)
4. Bouri, E., Shahzad, S.J.H., Roubaud, D.: Cryptocurrencies as hedges and safe-havens for US equity sectors. Q. Rev. Econ. Finance **75**, 294–307 (2020)
5. Canh, N.P., Wongchoti, U., Thanh, S.D., Thong, N.T.: Systematic risk in cryptocurrency market: evidence from DCC-MGARCH model. Financ. Res. Lett. **29**, 90–100 (2019)
6. Ciaian, P., Rajcaniova, M.: Virtual relationships: Short-and long-run evidence from BitCoin and altcoin markets. J. Int. Financ. Markets Instit. Money **52**, 173–195 (2018)
7. Conlon, T., Corbet, S., McGee, R.J.: Are cryptocurrencies a safe haven for equity markets? an international perspective from the COVID-19 pandemic. Res. Int. Bus. Financ., 101248 (2020)
8. Corbet, S., Lucey, B., Urquhart, A., Yarovaya, L.: Cryptocurrencies as a financial asset: A systematic analysis. Int. Rev. Financ. Analysis **62**, 182–199 (2019)
9. Corbet, S., Meegan, A., Larkin, C., Lucey, B., Yarovaya, L.: Exploring the dynamic relationships between cryptocurrencies and other financial assets. Econ. Lett. **165**, 28–34 (2018)
10. Corbet, S., Larkin, C., Lucey, B.: The contagion effects of the covid-19 pandemic: evidence from gold and cryptocurrencies. Financ. Res. Lett., 101554 (2020)
11. Engle, R.F., Granger, C.W.J.: Cointegration and error correction: representation, estimation, and testing. Econometrica **55**, 251–276 (1987)
12. Ferreira, P., Pereira, É.: Contagion effect in cryptocurrency market. J. Risk Financ. Manag. **12**(3), 115 (2019)
13. Goodell, J. W.: COVID-19 and finance: agendas for future research. Financ. Res. Lett., 101512 (2020)
14. Granger, C.W.J.: Developments in the study of cointegrated economic variables. Oxford Bull. Econ. Stat. **48**, 213–228 (1986)
15. Granger, C.W.: Some recent development in a concept of causality. J. Econmet. **39**(1–2), 199–211 (1988)
16. Iqbal, N., Fareed, Z., Wan, G., Shahzad, F.: Asymmetric nexus between COVID-19 outbreak in the world and cryptocurrency market. Int. Rev. Finan. Anal. **73**, 101613 (2021)
17. Ji, Q., Bouri, E., Lau, C.K.M., Roubaud, D.: Dynamic connectedness and integration in cryptocurrency markets. Int. Rev. Financ. Anal. **63**, 257–272 (2019)
18. Johansen, S., Juselius, K.: Maximum likelihood estimation and inference on cointegration with applications to the demand for money. Oxf. Bull. Econ. Stat. **52**, 169–210 (1990)
19. Karim, B.A., Majid, M.S.A.: Does trade matter for stock market integration? Stud. Econ. Finance **27**(1), 47–66 (2010)
20. Katsiampa, P., Corbet, S., Lucey, B.: Volatility spillover effects in leading cryptocurrencies: a BEKK-MGARCH analysis. Financ. Res. Lett. **29**, 68–74 (2019)
21. Kawohl, W., Nordt, C.: COVID-19, unemployment, and suicide. Lancet Psychiatry **7**(5), 389–390 (2020)
22. Kristoufek, L.: What are the main drivers of the Bitcoin price? Evidence wavelet coherence analysis. PloS One **10**(4), e0123923 (2015)

23. Lahmiri, S., Bekiros, S.: The impact of COVID-19 pandemic upon stability and sequential irregularity of equity and cryptocurrency markets. Chaos Solitons Fractals, 109936 (2020)
24. Liu, W.: Portfolio diversification across cryptocurrencies. Finance Res. Lett. **29**, 200–205 (2019)
25. Narayan, P.K., Smyth, R.: Higher education, real income and real investment in China: Evidence from Granger causality tests. Educ. Econ. **14**(1), 107–125 (2006)
26. Narayan, P., Smyth, R., Nandha, M.: Interdependence and dynamic linkages between the emerging stock markets of South Asia. J. Account. Financ. **44**, 419–439 (2004)
27. Pesaran, M.H., Shin, Y., Smith, R.: Bounds testing approaches to the analysis of level relationships. J. Appl. Econ. **16**, 289–326 (2001)
28. Pesaran, M.H., Shin, Y.: An autoregressive distributed lag modelling approach to cointegration analysis. DAE Working Paper No. 9514, Department of Applied Economics, University of Cambridge (1995)
29. Phan, D.H.B., Narayan, P.K.: Country responses and the reaction of the stock market to COVID-19—A preliminary exposition. Emerging Markets Financ. Trade **56**(10), 2138–2150 (2020)
30. Yarovaya, L., Matkovskyy, R., Jalan, A.: The COVID-19 Black Swan Crisis: Reaction and Recovery of Various Financial Markets (May 27, 2020). SSRN: https://ssrn.com/abstract=361 1587 or http://dx.doi.org/10.2139/ssrn.3611587
31. Yi, S., Xu, Z., Wang, G.J.: Volatility connectedness in the cryptocurrency market: is Bitcoin a dominant cryptocurrency? Int. Rev. Financ. Anal. **60**, 98–114 (2018)
32. Yousaf, I., Ali, S.: Discovering interlinkages between major cryptocurrencies using high-frequency data: new evidence from COVID-19 pandemic. Financ. Innovation **6**(1), 1–18 (2020)

Monetary Policy Shocks and the Term Structure of Bond Yield: A SVAR Analysis of Malaysia

Ahmad Khidir Othman, Zulkefly Abdul Karim, and Mohd Azlan Shah Zaidi

Abstract This chapter examines the impact of monetary policy shocks (domestic and foreign) and oil price shocks on bond yield movement across the term structure. Indicators of monetary policies used were overnight interbank rate (IBOR) for domestic monetary policy and federal funds rate (FFR) for foreign monetary policy. In contrast, bond yield data was from the index of the bond yield of three−year, five−year, and ten−year maturity rates. The exogenous shocks have generated using structural vector autoregressive (SVAR) in an open economy approach. The monthly data spanning from June 1998 until January 2019 has been used in examining the propagation of the monetary policy and oil price impulses on the movement of public and private bonds across maturity periods. The main findings using the impulse response function revealed that monetary policy shocks contribute to bond yield's positive movement across the term structure. However, the oil price shock causes a negative direction of the bond yields.

Keywords Monetary policy shocks · Bond yields · SVAR · Overnight interbank rate · FFR · Oil prices · Impulse response · Variance decomposition · Malaysia · Public and private bond

1 Introduction

The question of the effect that monetary policy has on the economy has been intriguing since the early 1900s. One of the most initial studies on this topic was written by [11] that focused on the history of money supply in the United States. This book changed the paradigm that typical Keynesian economists had about how monetary policy has little influence on the economy. This book showed how the adjustment of the money supply by the Federal Reserve through the channel of monetary policy affected the growth of nominal income, real output, and price level.

A. K. Othman · Z. A. Karim (✉) · M. A. S. Zaidi
Center for Sustainable and Inclusive Development (SID), Faculty of Economics and Management, The National University of Malaysia (UKM), 43600 Bangi, Selangor, Malaysia
e-mail: zak1972@ukm.edu.my

© Institute of Technology PETRONAS Sdn Bhd 2022
S. A. Abdul Karim (eds.), *Shifting Economic, Financial and Banking Paradigm*,
Studies in Systems, Decision and Control 382,
https://doi.org/10.1007/978-3-030-79610-5_6

When a country adjusts its monetary policy, the main goal is always to ensure the economy's sustainable growth. The adjustment might occur by altering the amount of money supply or changing the interest rate on overnight loans. These adjustments will stimulate a slow economy or prevent a bubbling economy from crashing. On the other hand, these adjustments will change securities' yields and, ultimately, their demand in the bond market. This can be explained through various premises, one being the expectations hypothesis. According to this hypothesis, it is assumed that a long-term bond yield is determined so that successive short-term current and future bond rates will be of equal value to that of the long-term bond. In this case, an increase in overnight lending will cause an upward movement in bond yield, provide that arbitrage opportunity between lending maturities is minimal. For example, a one-year bond yield will result from the return of current and the expected overnight loans' rate for 365 days.

Keynes' liquidity preference theory can also explain the movement of yield in the bond market due to changes in monetary policy. According to this theory, individuals demand liquidity for transactions as a precautionary and speculative motive. As a result, to entice individuals to relinquish their liquid money for investment purposes, a return is required. The asset return should be more prominent as time increases since individuals relinquish their money for a longer time horizon. Longer maturity bonds will also be exposed to higher risks due to more uncertainties that can occur within a longer time horizon and result in a higher yield.

Figure 1 shows the relationship between short-term and long-term bonds, also known as the term structure of bond yield. Figure 2 shows the Overnight Policy Rate (OPR) impact and Federal Fund Rate (FFR) on a 3-year government bond yield. As shown in Fig. 2, the government 3-year bond yield always stays above OPR, and OPR is somewhat affected by the movement of FFR due to Malaysia being a small open economy.

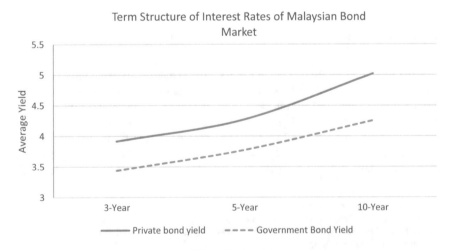

Fig. 1 Term structure of interest rates of Malaysian bond market

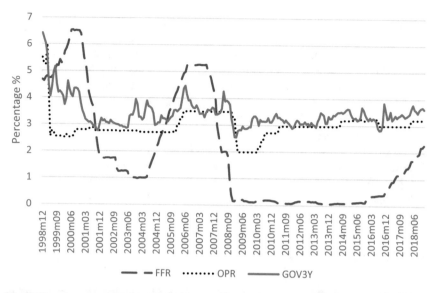

Fig. 2 The movement of Federal Fund Rate (FFR), Overnight Policy Rate (OPR), and three years of Malaysian Government Bond (GOV3Y)

In Malaysia, monetary policy is controlled by the Central Bank of Malaysia (BNM) by the overnight policy rate mechanism. This mechanism was employed since 2004 and was being cited as being more transparent and enabling market participants to make better decisions. However, since the adoption of OPR in 2004, BNM has only adjusted the rate fourteen times. Most of these changes occurred due to economic activity changes, as seen during the global financial crisis when BNM dropped the rate three times, from 2008 until 2009, and later raised the rate when the economy started to recover. As mentioned above, the changes in OPR will result in changes in bond market returns. Therefore, it is an exciting topic to study the impact of these changes, specifically the propagation of monetary policy shocks towards the term structure of bond yield in Malaysia. This study is essential for policymakers to control the number of activities in the Malaysian bond market. It will also be necessary for the market participant to anticipate market movements from monetary policy shock.

This study attempts to contribute to the existing research on the effects of monetary policy changes on bond yield in several ways. First, this paper studies the impact of monetary policy shocks on the term structure of bond yield from the Malaysian bond market's perspective. Although [6, 16] have examined the determinants of the Malaysian bond market, their study not focused on monetary policy shocks upon the term structure of the bond market. Despite not being a new area to focus on, there are already various papers discussing this issue, mainly from the United States (U.S) and other developed nations,however, only a few studies have been conducted in developing nations. Therefore, this present study's focal point is to examine this issue from a developing country's perspective. Lastly, this study has augmented the standard exogenous shock of monetary policy model from [7] by adding variables that might affect the domestic bonds, namely the Federal Funds Rate, to better isolate the effect of foreign monetary policy shock to domestic bond yields. Most existing studies do not incorporate foreign monetary policies since the study's focus is on developed countries where foreign monetary policies have no impact on their bond market. Thus, we are trying to contribute to the existing research by studying the impact foreign monetary policies have on countries with a small open economic structure like Malaysia.

The structure for the rest of this paper is as follows. Section 2 discusses previous literature on this topic, whereas, Sect. 3 describes the econometric model of SVAR and the justification of selecting the model. Section 4 reports the main findings, and Sect. 5 concludes.

2 Literature Review

The study between monetary policy shocks and bond yield has been intriguing in finance and macroeconomics since the early 1990s. Various studies focused on this topic, such as [3, 9, 17]. This study became the main interest of many people due to its impact on the economy. By predicting the movement of yield in response to shocks, various counter-measures and prospective actions can be taken, especially for policymakers and bond market participants.

Early studies from [11], and supported by [18], and, etc., have found that monetary policy actions cause a reaction in the real output of a country that may last for a specific period. Despite many studies on the impact of monetary policy on economic variables, they could not find a consensus on the transmission of the impact. The conventional wisdom of the transmission is,whenever a monetary policy is adjusted using short-term interest rates, the cost of capital will be affected tremendously and, consequently, affecting expenditure on real assets, houses, and inventories, as well as investment products. [4] argued that this conventional wisdom is not strongly supported by empirical evidence. The proposed monetary policy transmission through a credit channel in which whenever a monetary policy is adjusted, the impact it has is mostly on the balance sheet of borrowers, causing the cost of financing to move and affect their market strategy. Monetary policy shock is influencing the external finance premium, thus affecting credit market participants' behavior.

As for the impact of monetary policy shocks on the term structure of bond yield, many angles have been explored, such as the paper from [15] that looked into Federal Funds' future market to study the impact of anticipated and unanticipated changes of monetary policy on bond rates. The result discovered no response in rates for anticipated changes and a highly significant and large response for unanticipated monetary policy changes. Papers from [5, 7, 12, 19], etc. meanwhile used the SVAR model to generate shocks in the economy and observe the impulse response it has on economic variables.

In the paper by [7], the main focus is to identify the monetary policy and identification restrictions. One of the identification restrictions used in the paper is benchmark policy ordering, using the assumption that central banks' decision on monetary policy is from the state of real economic activity. [7] employs a recursive identification strategy that differs from [19] that used a nonrecursive identification strategy and [12] who utilized long run and contemporaneous restrictions on monetary policy shocks. These papers focused mainly on the discussion and justifications for their identification strategy to identify the channel in which monetary policy transmission is transferred into the economy.

Taking their ideas into consideration [10], used all three identification strategies and investigated the impact of monetary policy shock upon the term structure of nominal interest rates specifically. Their study used risk-free bonds as the proxy for the term structure of interest rates and monthly data of Federal Fund rates, Producer Price Index, Inflation, sensitive material prices index, and growth rate of monetary aggregate M2. The main findings using a different identification strategy revealed that the impulse response and variance decomposition closely resemble each other. They found that monetary contraction, as expected, raises the interest rates of bonds. The effect is temporary, which occurs between six to twelve months after the initial shock. The shock also primarily affects short-term rates and, on a lesser scale, long-term rates, with most of the response of long-term rate bonds is explained through the expectation hypothesis.

Another paper that discusses the same topic is by [8], with the difference being that it focuses solely on the identification strategy brought up by Christiano. The variables used are similar to [10], except for the exemption of the impact coupon payment might have on determining the yield. The results from the study are also similar to [10], where there is a substantial response of bonds with shorter-term maturities on exogenous monetary policy shock, and the effect is only temporary. Again, longer-term bond yields' response can be explained by the expectations hypothesis, whereby the movement of shorter-term bonds affects the variability of longer-term bond yields. This causes a weaker, smaller magnitude response of longer-term bonds to an exogenous monetary shock. The paper also found the impact of material prices on long-term yields.

For literature regarding the influence of foreign monetary policy towards domestic bond markets, [1] found a significant impact of the U.S monetary policy changes towards domestic bond markets of various countries globally, both developed and emerging market economies. The result is statistically significant, with the impact's magnitude comparable to events affecting bond yields, including domestic monetary policy shocks. Meanwhile, the influence oil price shocks have on the economy was discussed by [2] and [14]. [14] paid closer attention to the impact of oil price shocks on the U.S bond market returns. They found that positive oil demand shocks, from market-specific simulation and positive innovation in the global aggregate, cause a significant decline in real bond index returns. Structural oil market shocks also have an essential predictive ability for real bond returns in the U.S bond market. The reason oil shock impacts bond yield movement is not straightforward. From the paper, oil shocks significantly affect the stock market, and since bonds and stocks are substitutes, the impact oil shock has on the stock market also affects the bond market in a reverse manner.

In the context of Malaysia, there have been several studies on the impact of monetary policy on the economy and the study on the bond market. For example, Zaidi and Fisher [20] studied the impact of monetary policy shock on output, inflation, and real exchange rate, while Zaidi et al. [21] studied the impact monetary policy shock has on disaggregated inflation. Hadi et al. (2019) discussed bond rates determinant for short-term and long-term securities using the error correction model (ECM) and wavelet analysis. Other papers that study the bond market is by Che-Yahya et al. [6], which focused on the selected determinants of corporate bond yield and concentrates on the individual issue level, and Che-Yahya et al. [6] that studied factors influencing yield spreads of Malaysian bonds using the multifactor model. A recent study by Mazwinda et al. [16] investigated the determinants of government bond yields and found that fiscal deficits, government debt, and exchange rates have significantly influenced government bond yields. However, the study does not focus on the role of monetary policy shocks on bond yields. Therefore, as far as we know, no literature explores the impact of monetary policy shocks on the term structure of bond yield in Malaysia. Accordingly, this study fills that gap of knowledge by investigating the behavior of rates of return of government and corporate bonds across terms of structure in response to domestic and foreign monetary policy and oil price shocks.

3 Methodology

This section will be discussing the methodology used in the study to see the relationship between monetary policy changes and bond yields in Malaysia. For this purpose, we used the Structural Vector Autoregression (SVAR) to study the impact of exogenous shocks on the bond yield of various maturities and issuances. We used monthly data from the year 1998 until 2018 for a better understanding of how the domestic and international monetary policy shocks transmit to the term structure of bonds yield. Many studies have been conducted specifically in identifying economic impulses using VAR methods. For example, Christiano et al. [7] used recursive strategy, whereas, [19] used a nonrecursive strategy, and [12] used a mixture of contemporaneous and long-run restrictions. This study uses the identification strategy proposed by CEE due to its sensibility in the case of small-open economies.

3.1 Monetary Policy Rules

From CEE, they underlined the basic structure for exogenous policy shocks this way. For central banks to adjust their monetary policy, they usually will look at several indicators in the economy that suggest that a change is necessary. We will refer to these indicators that affect the policy maker's actions as their *feedback rule*. However, it is normal that policymakers also adjust their policy following changes that are not part of the feedback rule, and we will call this the exogenous monetary policy shock. So, we can simplify the changes in monetary policy actions due to the feedback rule and exogenous shock.

With that, we base our model as.

$$OPR_t = f(\Omega_t) + \varepsilon. \tag{1}$$

In the above equation, Ω_t is our feedback rules, in which the set of variables that the central back has available at time t, f is a linear function that defines the central bank's reaction to feedback rules, and ϵ is our exogenous shock to monitory policy. The policy reaction function f is different from country to country, but the general rule applies to all in which monetary authority will prefer policy-making actions to stabilize the economy, control inflation, and many more. The residual ϵ is arbitrary, nonsystematic factors that depend on the personalities and opinions of the Monetary Policy Committee.

To construct the exogenous shocks of a monetary policy, we adjusted CEE's list of variables to suit the availability of data in Malaysia and the context of a small open economy. We included foreign monetary policy and the GDP as well. In CEE model, they used real GDP, GDP deflator, non-borrowed reserves, total reserves, federal fund rate and commodity prices. But to suit our context of research, we changed real GDP to Industrial Price Index (IPI) as an indicator for real economic activity, CPI inflation as a substitute for GDP deflator (INF), excluded the usage of total reserves and non-borrowed reserve and changed it to money supply (M3), kept Federal Funds Rate (FF) and introduced Overnight Policy Rate (OPR) and governments and private bond yield of maturity 3, 5, and 10 years as the interest of our study.

We can categorize our variables into several types. First is the foreign variables. In this case, it is the Federal Funds Rate (FFR) as foreign monetary policy, and we also included USGDP, the GDP of the United States, as it influences the Federal Funds Rate. The second type is domestic variables, data that our country produces. Those are MYGDP, INF, M3, and OPR. The third type is our research interest is the yield of government and corporate bonds of the maturities 3, 5, and 10 years. For the most part, it is better to take the yield of zero-coupon bonds due to having no coupon payment associated with the bonds that might affect the yield of the bonds, but due to the lack of data, normal bonds were used. The last variable is the price of the commodity. In this case, we used oil due to Malaysia's dependency towards oil. All the data are taken from Bloomberg and Reuters spanning from the beginning of 2000 until the end of 2018 (18 years or 216 months).

Thus, the model has become: OIL, USGDP, FFR, MYGDP, INF, M3, OPR, and Y.

First, we have to estimate the baseline SVAR model. It can be written as follows:

$$A_0 Y_t = \Gamma_0 D_0 + A(L) Y_t + \varepsilon_t \tag{2}$$

Where A_0 is a square matrix which shows the coefficient that interacts directly in the structure between the variables in the system, Y_t is an (8×1) matrix that is the vector of system variables or $(log OIL, log USGDP, FFR, log MYGDP, inf, \Delta_{m3}, OPR, and_Y)$, D_0 is a vector of deterministic variables, (constant, trend and dummy variables), A(L) is a *kth* order matrix polynomial in the lag operator $L[A(L) = [A_1 L - A_2 L - \ldots - A_k L^k]$, and ε_t is an (8×1) vector structural disturbance. Equation (2) cannot bed estimated directly to obtain the value of A_0. However, the identification of matrix A_0 can be estimated by reduced form as follow:

$$Y_t = A_0^{-1} \Gamma_0 D_0 + A_0^{-1} A(L) Y_t + A_0^{-1} \varepsilon_t \tag{3}$$

or.

$$Y_t = \Pi_0 D_0 + \Pi_1(L) Y_t + \mu_t \tag{4}$$

where.

$$\Pi_0 = A_0^{-1} \Gamma_0, \Pi_1 = A_0^{-1} A(L) Y_t, \mu_t = A_0^{-1} \varepsilon_t. \tag{5}$$

This study used the A-structured VAR model to create identifiers among current parameters (contemporaneous parameter). Thus, the identification scheme of the contemporaneous matrix in SVAR model is stated as the following matrix:

$$\begin{pmatrix} 1 & 0 & 0 & 0 & 0 & 0 & 0 & 0 \\ a_{21} & 1 & 0 & 0 & 0 & 0 & 0 & 0 \\ a_{31} & a_{32} & 1 & 0 & 0 & 0 & 0 & 0 \\ a_{41} & 0 & 0 & 1 & 0 & 0 & 0 & 0 \\ a_{51} & 0 & 0 & a_{54} & 1 & 0 & 0 & 0 \\ a_{61} & a_{62} & a_{63} & a_{64} & a_{65} & 1 & 0 & 0 \\ 0 & 0 & 0 & a_{74} & a_{75} & a_{76} & 1 & 0 \\ a_{81} & 0 & 0 & a_{84} & a_{85} & a_{86} & a_{87} & 1 \end{pmatrix} \begin{bmatrix} u_{oil} \\ u_{usgdp} \\ u_{ffr} \\ u_{mygdp} \\ u_{inf} \\ u_{opr} \\ u_{\Delta M3} \\ u_{yield} \end{bmatrix} = \begin{bmatrix} e_{oil} \\ e_{usgdp} \\ e_{ffr} \\ e_{mygdp} \\ e_{inf} \\ e_{opr} \\ e_{\Delta M3} \\ e_{yield} \end{bmatrix} \tag{6}$$

where $e_{oil}, e_{usgdp}, e_{ffr}, e_{mygdp}, e_{inf}, e_{\Delta M3}, e_{opr}$ and e_{yield} are structural disturbances or structural shocks of oil price, foreign output, foreign monetary policy, domestic output, domestic inflation, domestic changes in money supply, domestic monetary policy and bond yield of various maturities. In the meantime, $u_{oil}, u_{usgdp}, u_{ffr}, u_{mygdp}, u_{inf}, u_{\Delta M3}, u_{opr},$ and u_{yield} are respective residual shock of the variables.

3.2   Explanation of the Restrictions

The matrix in Eq. 6 shows that world oil price has been used to proxy the adverse supply shocks in exchange for commodity prices that have been used by the previous study, for example, [7, 12, 19]. This is due to the nature of the two variables being similar thus exchangeable, besides the fact that oil price as a proxy is more suitable in the context of Malaysia. In this model, oil prices (OIL), foreign output (USGDP), and the federal funds rate are variables that are unaffected by changes that happen domestically. Gross domestic product of the U.S (USGDP) is assumed only affect foreign and domestic monetary policy in our model, which is Federal Funds Rate and Overnight Policy Rate respectively, since the Federal Reserve of the United States will use output of the country to forecast the size of inflation. The Central Bank of Malaysia will also use its monetary policy rate adjustment as a benchmark for the domestic monetary policy. These will result in FFR and OPR being adjusted accordingly.

Domestically, the Malaysian GDP responds contemporaneously only by the changes in oil prices if we assume that price level, domestic interest rates, exchange rates, and other variables do not directly affect foreign output. As is generally known, the growth in domestic production is highly dependent and moves in tandem with oil prices. The domestic price level is influenced by oil prices and the Malaysian GDP or output. An increase in world oil prices will increase the cost of various products, whether directly or indirectly related to the oil and gas industry. According to Fisher's equation, inflation is also affected by output, states that a drop in production will increase inflation. The domestic monetary policy equation has already been discussed at the beginning of this section at Eq. (1), where the central bank will adjust monetary policy according to feedback rules. In this case, our feedback rule is oil price, domestic output, and inflation. The change in money supply is a standard specification for the money demand equation that depends on domestic output, price level, interest rate, and monetary policy. Domestic monetary policy is controlled to ensure that economic growth can be sustained in the long run. The last variable is the bond yield of various maturities that are influenced contemporaneously by changes of all variables except for foreign output and foreign monetary policy.

By basing our matrix from (6), the VAR model can form 8 residual equations. After we estimate the baseline SVAR model, the accumulated response of economic growth following the monetary shock is constructed. The structural shocks can be identified by the calculation of the different variables to shocks in monetary policy instruments which are also known as impulse response function (IRF). This IRF is a traditional feature of the response to monetary policy. The SVAR was ran using 229 observations (18 years of monthly data).

3.3 *Impulse Response Function*

As per the usual procedure of the SVAR model, we will be interested in looking at the impulse response function of a variable of interest that comes as a result of shock from changes to endogenous variable in the model. A surprise of the immediate endogenous variables affects its self and other variables inside the structural VAR model.

Conceptually, it can be described as:

$$y_{1t} = a_{11}y_{1t-1} + a_{12}y_{2t-1} + \varepsilon_{1t} \tag{7}$$

$$y_{2t} = a_{21}y_{2t-1} + a_{22}y_{2t-1} + \varepsilon_{2t} \tag{8}$$

At time t, the error term ε_{1t}, which represents shocks, directly affects y_{1t}, but no direct effect on y_{2t}. In the meantime, at time $t + 1$, the shock on y_{1t} will continually cascade affecting y_{1t-1} and y_{2t-1}. The shock will continue throughout the period $t + 2$, $t + 3$, and thereupon. This shock will form a chain reaction to all variables incorporated in the model, and this chain reaction is called the impulse response function.

4 Empirical Results

As stated in the introduction, this paper aims to examine the impact of domestic and foreign monetary policy shocks on the yield of bonds of various maturities. We are also interested in looking at other shocks affecting bond yield, such as oil shocks.

4.1 *Impulse Response*

As mentioned above, the subject of interest in constructing an SVAR model is to obtain the impulse response function and see the reaction of interest in the shock of certain variables. Therefore, in Fig. 3, we plotted the estimated response of Malaysian bond yields to a one-standard-deviation contractionary domestic monetary policy shock. Bond yields are taken from the index available; 3-year government bond yield index, 5-year government bond yield index, 10-year government bond yield index, 3-year corporate bond yield index, 5-year corporate bond yield index, and 10-year corporate bond yield index. The yields are monthly data from December 1998 until December 2018.

The solid blue lines of the plot are the estimation of the response; the red lines above and below it act as boundaries of the 95% confidence region. The plots trace the response over ten months and are measured in percent deviation from

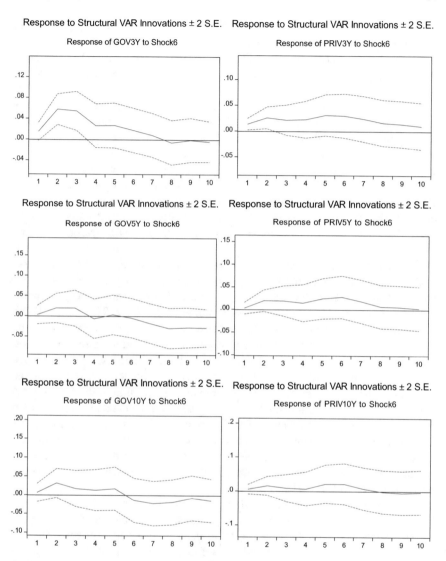

Fig. 3 Response of bond yield to domestic monetary policy shocks

the nonstochastic steady state, with one standard deviation amounted to around 50 basis points. The displayed plots respond to governments and private bonds with maturities of three years, five years, and ten years.

A one standard deviation domestic monetary policy shocks, or 50 basis point, increases the one-year government bond yield to 5 basis points in the first month, and it climbs to around 30 basis points at its peak, which occurs during the second month after the shock happened. The response is significant until the fourth month after the shock, in which the response starts to diminish after the seventh month. As for the one-year private bond yield, the increase post domestic monetary policy shock is similar to the government counterpart bond, at five basis points, but it experiences a lower initial peak, at around ten basis points during the second month. Later during the fifth and sixth months, the second higher peak appears, clocking around 15 basis points. The response is also statistically insignificant after the second month.

The response of government five-year and ten-year bond yield encounters similar response patterns. Both peaked during the second month at around the same basis point of 10. The response of government 5-year bond yield however goes toward zero earlier, during the third month, than 10-year bond yield, at fifth month. Both yields are statistically insignificant. Their counterpart at the private bonds also has a similar pattern. Interestingly, this response pattern is consistent with various literature, [8, 10], where both papers have their bond yield to peak in response to shocks. The peak occurs during the second-month horizon. Additionally, their result also showed that longer-bond terms give smaller impulse response, and all the impulses dissipated as time moves forward.

We also look at the response bond yield has to foreign monetary policy shock in Fig. 4. An increase in foreign monetary policy shock of about 50 basis points will result in a hike up to about 30 basis point increase in three-year government bond yield at its peak during the fourth month after the shock. The yield will later drop and climb up again, giving the plot two peaks across its horizon. A similar pattern has been observed in the five-year government bond yield response, albeit a lower magnitude of basis point movements; only 20 and 25 basis points for five-year and ten-year government bond yield, respectively. The three-year and ten-year government bond yield reactions are statistically significant until the fourth month but are not statistically significant for the five-year government bond. Its counterpart, the yield of private bonds, meanwhile, all give a non-significant impulse response. Despite that, they all provide the same response pattern, and again the only difference being the magnitude of the response. They all peak at the fourth month after the shock, at around ten basis points for 3-year and 15 basis points for 5-year bonds and 10-year bonds at approximately 30 basis points.

Lastly, in Fig. 5, we look at how a shock on oil price affects the yield of Malaysian bonds. As usual, the response pattern for all government bond yields is the same, with the difference being the time for the response to dissipate. The response for a longer-maturity term bond dissipated quicker than a shorter-maturity term bond, at the fourth months after the oil shock for 10-year and 5-year bond and around the fifth month for the 3-year government bond. The basis point is the highest, as usual, for a three-year government bond at around ten basis points, while the longer-term government bonds hover below ten basis points at their peak. In the meantime, the response for all private bonds seems to be different, unlike other shocks. The three-year private bond started at a negative value and rose to a little bit above positive value in the seventh month. A five-year private bond never reached positive value for the entirety of the time horizon. As for the ten-year private bond, the response reaches zero in the second and fourth months after the oil shock occurs, but never higher than that.

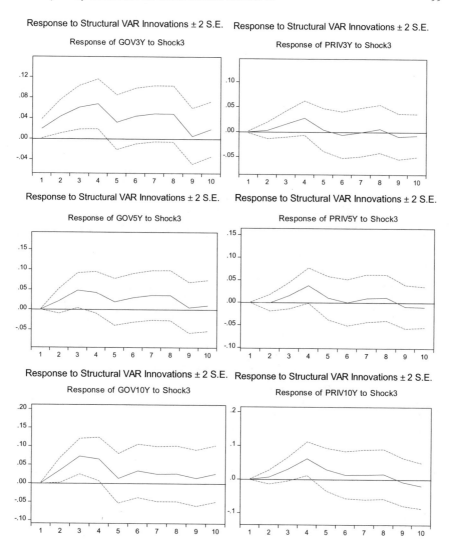

Fig. 4 Response of bond yield to foreign monetary policy shock

4.2 Variance Decomposition

From the impulse response function, we can propose that monetary policies, domestic and foreign, play a decisive role in determining the variability of a short-run interest rate for bonds with short-term maturity. To discuss this point further, we look at variance decompositions in Table 1 that estimate the conditional variance of yield of bonds with various maturities attributable to domestic monetary-policy shock,

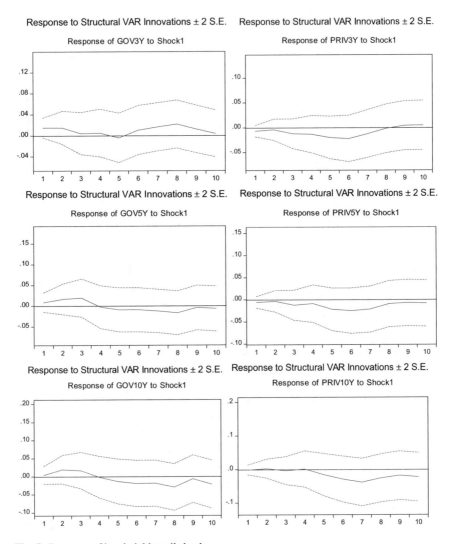

Fig. 5 Response of bond yield to oil shock

foreign monetary-policy, and oil price shocks. The variance decomposition has also been identified at 1-month, 6-month, 12-month, and 24-month time horizons.

From the table, we identify that the shorter the term of maturity, the stronger the influence of shocks on its variability. Surprisingly, we see a relatively more substantial impact on Malaysian bond yield by foreign monetary policy shock than domestic monetary policy shock. According to Kearns et al. (2018), this is due to three factors; (1) Tighter economic relation between domestic and foreign countries will create a bigger spillover due to shock. (2) Second is the channel in which foreign money exchange is managed in a particular country. Pegging domestic currency to foreign

Table 1 Fraction of yield variance explained by

A. *Domestic Monetary Policy Shock*

Horizon	3-year Govt Bond (%)	5-year Govt Bond (%)	10-year Govt Bond (%)	3-year Private Bond (%)	5-year Private Bond (%)	10-year Private Bond (%)
1 month	1.1	0.1	0.1	2.7	0.09	0.2
6 months	8.1	0.6	1.1	5.1	2.6	0.8
12 months	7.2	2.1	1.2	5.3	2.7	0.6
24 months	6.17	1.8	1.0	5.2	2.5	0.5

B. *Foreign Monetary Policy Shock*

Horizon	3-year Govt Bond (%)	5-year Govt Bond (%)	10-year Govt Bond (%)	3-year Private Bond (%)	5-year Private Bond (%)	10-year Private Bond (%)
1 month	1.8	0.01	0.001	0.008	0.001	0.0007
6 months	13.1	4.2	6.3	1.4	1.9	3.6
12 months	15.1	4.9	5.7	1.3	2.0	3.2
24 months	14.1	5.1	4.9	1.1	1.9	2.8

C. *Oil PriceShock*

Horizon	3-year Govt Bond (%)	5-year Govt Bond (%)	10-year Govt Bond (%)	3-year Private Bond (%)	5-year Private Bond (%)	10-year Private Bond (%)
1 month	1.1	0.2	0.01	0.7	0.4	0.01
6 months	0.5	0.6	0.6	1.7	1.6	0.7
12 months	1.2	0.8	1.2	1.6	2.0	0.2
24 months	1.8	0.7	1	1.3	1.9	0.1

currency will experience a bigger influence in shock. (3) Lastly, the risk premium channel, whereby a country which focuses on attracting foreign investor, by being more open, will be strongly affected by foreign monetary policy shock.

As for the impact of oil price shock on the term structure of bond yield, we see a steady influence of shocks on bonds of various maturities up until the 24-month time horizon. Additionally, the impact later dissipates as time moves forward, according to a previous study from [14] that also have the same result.

Alas, we can concede to the fact that at a one-month horizon, for all bond maturities, the forecast error variance is attributed by the yield's own shock (error term in the matrix); around 90% for three-year government and private bond, approximately 95% for five-year and ten-year government and private bond yield.

5 Summary and Conclusion

The standard views on monetary policy transmission mechanism are based on a reliable relationship between monetary policy actions and market interest rates. The empirical result shows that monetary policy shock, whether it comes from within the country or outside, influencing the yield of bonds in Malaysia. The movement of oil price has also essential in determining the level of bond yield in Malaysia. While there is considerable evidence that monetary policy has a significant impact on short-term interest rates, there is often a weaker and less reliable connection between policy actions and long-term rates. Our main finding found out that with monetary policy shock, all bonds will rise in yields. It primarily affects shorter-term bonds, with longer-term rates are influenced but with a smaller magnitude. In addition to that, the study found out that both monetary policies, domestic or foreign, play a significant role in determining the variability of the yield. We also found that the global price of oil plays a somewhat important role in determining the return of a bond.

We believe that these findings can be an addition to the body of knowledge in the context of the Malaysian bond market movement and its relationship with monetary policy shocks. It is also beneficial to market players to look at the preference of monetary authorities in determining monetary policy due to its impact on the bond market. As for monetary authority, the findings from this study show that monetary policy affects bond market returns. These findings indicate that monetary policy via interest rates policy can be a considerable tool to control the Malaysian bond market. With this knowledge, the Central Bank of Malaysia would adjust monetary policy accordingly to stabilize the domestic bond market. Market participants, consequently, can observe monetary authority's preference on market activities to predict their next policy order so that they can plan their investment strategy accordingly. The foreign factors, namely world oil price and foreign monetary policy, also need to be observed closely by the monetary authority and the market participant. Both exogenous shocks are statistically significant in influencing the domestic bond market; therefore, it is crucial to understand how the foreign shocks have transmitted to the local bond market for monetary policy formulation and portfolio management.

Acknowledgements We would like to thanks Dr. Mohd Afzanizam Abdul Rashid, Chief Economist of Bank Islam Malaysia Berhad, for providing the data, and all participant of the Post Graduate Colloquium organized by the Faculty of Economics and Management, Universiti Kebangsaan Malaysia (UKM), Bangi, Selangor, Malaysia for their valuable comments and input.

References

1. Albagli, E., Ceballos, L., Claro, S., Romero, D.: Channels of US monetary policy spillovers to international bond markets. Bank for International Settlements Working Paper, No. 719 (2018)
2. Bernanke, B.S., Gertler, M., Watson, M.: Systematic monetary policy and the effects of oil price shocks. Brook. Pap. Econ. Act. **1997**(1), 91–157 (1997)
3. Bernanke, B.S., Blinder, A.S.: Credit, money, and aggregate demand. Am. Econ. Rev. **78**(2), 435–439 (1988)
4. Bernanke, B.S., Gertler, M.: Inside the black box: the credit channel of monetary policy transmission. J. Econ. Perspect. **9**(4), 27–48 (1995)
5. Bernanke, B.S.: Alternative explanations of the money-income correlation. In: Carnegie-Rochester Conference Series on Public Policy, vol. 25, pp. 49–99 (1986)
6. Che-Yahya, N., Abdul-Rahim, R., Mohd-Rashid, R.: Determinants of corporate bond yield: the case of Malaysian bond market. Int. J. Bus. Soc. **17**(2), 245–248 (2016)
7. Christiano, L., Eichenbaum, M., Evans, C.: Identification and the Effects of Monetary Policy Shocks. Financial Factors in Economic Stabilization and Growth, Cambridge University Press (1996)

8. Edelberg, W., Marshall, D.: Monetary policy shocks and long-term interest rates. Econ. Perspect. **20**(2), 1–17 (1996)
9. Estrella, A., Hardouvelis, G.A.: The term structure as a predictor of real economic activity. J. Financ. **XLVI**(2), 555–576 (1991)
10. Evans, C.L., Marshall, D.A.: Monetary policy and the term structure of nominal interest rates: evidence and theory. Carn. Roch. Conf. Ser. Public Policy **49**, 53–111 (1998)
11. Friedman, M., Schwartz, A.J.: A Monetary History of The United States, 1867–1960. Princeton University Press, N.J (1963)
12. Gali, J.: How well does the is-lm model fit post war data? Q. J. Econ. **107**(2), 709–738 (1992)
13. Hadi, A., Razak, A., Zainuddin, Z., Hussain, H.I., Rehan, R.: Interactions of short-term and long-term interest rates in Malaysian debt markets: application of error correction model and wavelet analysis. Asian Acad Manag J **24**, 19–31 (2019)
14. Kang, W., Ratti, R.A., Kyung, H.Y.: The impact of oil price shocks on U.S. bond market returns. Energy Eco. **44**, 248–258 (2014)
15. Kuttner, K.: Monetary policy surprises and interest rates: evidence from the fed funds futures market. J. Monet. Econ. **47**(3), 523–544 (2000)
16. Mazwinda, M., Karim, Z.A., Karim, B.A.: Penentu pergerakan hasil bon kerajaan di Malaysia: analisis model auto regresif lat tertabur (ARDL). Asian J. Account. Gov. **12**, 25–35 (2019)
17. Mishkin, F.S.: What does the term structure tell us about future inflation. J. Monetary Econ. **25**(1), 77–95 (1990)
18. Romer, C.D., Romer, D.H.: Does monetary policy matter? a new test in the spirit of Friedman and Schwartz. NBER Macroecon. Annual **4**, 121–184 (1989)
19. Sims, C., Zha, T.: Does monetary policy generate recessions? Macroecon. Dyn. **10**, 231–272 (1995)
20. Zaidi, M.A.S., Fisher, L.A.: Monetary policy and foreign shocks: a SVAR analysis for Malaysia. Korea World Econ. **11**(3), 527–550 (2010)
21. Zaidi, M.A.S., Karim, Z.A., Azman-Saini, W.N.W.: Relative price effects of monetary policy in Malaysia: A SVAR analysis. Int. J. Bus. Soc. **17**(1), 47–62 (2016)

Assessing Consumers' Preferences in Purchasing Green Vehicles: An Economic Valuation Approach

Hanny Zurina Hamzah and Sharul Shahida Shakrein Safian

Abstract Environmental degradation has been a challenge to everyone globally. Despite the burgeoning concern of the current environmental issues, there has been an increased level of awareness regarding environmental issues based on the worsening situation in developing countries. Therefore, this study investigates the preferences of Malaysians in using green vehicles through choice modelling (CM), using choice experiment of attributes: Level of Emissions, Range of Driving, Source of Energy and Price. In achieving this objective, data was collected from a sample of 427 respondents through survey questionnaires which were personally distributed by the researcher. The questionnaires contain information about green vehicles attributes and the data gathered were analysed using Economic Valuation Method. The result shows that the highest marginal value from range of driving is 20.5%, which suggests that people are concerned about their driving range; which is an indication of their preferences for green vehicles.

Keywords Economic valuation · Automotive industry · Green vehicle · Consumer preferences

Environmental degradation has been a challenge to everyone globally. Despite the burgeoning concern of the current environmental issues, there has been an increased level of awareness regarding environmental issues based on the worsening situation in developing countries. Therefore, this study investigates the preferences of Malaysians in using green vehicles through choice modelling (CM), using choice experiment of attributes: Level of Emissions, Range of Driving, Source of Energy and Price. In achieving this objective, data was collected from a sample of 427 respondents through questionnaires which were personally distributed by the researcher.

H. Z. Hamzah (✉)
School of Business and Economics, Universiti Putra Malaysia, 43400 Serdang, Malaysia
e-mail: hannyzurina@upm.edu.my

S. S. S. Safian
Department of Economics and Financial Studies, Faculty of Business Magement, Universiti Teknologi Mara (UiTM), Puncak Alam Campus, 42300 Selangor, Malaysia

© Institute of Technology PETRONAS Sdn Bhd 2022
S. A. Abdul Karim (eds.), *Shifting Economic, Financial and Banking Paradigm*,
Studies in Systems, Decision and Control 382,
https://doi.org/10.1007/978-3-030-79610-5_7

The questionnaires contain information about green vehicles attributes and the data gathered were analysed using Economic Valuation Method. The result shows that the highest marginal value from range of driving is 20.5%, which suggests that people are concerned about their driving range; which is an indication of their preferences for green vehicles.

1 Introduction

Previous studies on environmental economics underpin the fact that green vehicles fundamentally release less emission to the environment than conventional vehicles (Sampaio et al. [14]). There is information asymmetry between sellers and buyers when buyers want to pick green vehicles over conventional ones due to the absence of knowledge and information about the green vehicles. As a result, there is a static sale of green vehicles throughout the years.

Table 1 shows the total green vehicle sales of six car models in Malaysia for the period of 2013–2015. The table illustrates that the percentage of sales had decreased by 11.7% from 2013 to 2014. Similar decrease in green vehicle sales is also seen in the period 2014–2015, where the sales reduced by 44.5% higher than the previous period. Although other competitors could hardly make a good business in the period of 2013–2014, Honda topped the green vehicles sales performance by selling 676 additional units (19%) in the same period. Despite a significant drop in overall green vehicle sales for the year 2014 and 2015, the green vehicles sales performance in the following period witnessed a huge success recorded by Nissan and Toyota by registering 1,328 units from 1,079 units and 1,996 units from 502 units, respectively.

During the first six months of 2014, green vehicle sales fell to 11.7%, with a total of 6,007 units from 6,803 units in the previous year (refer to Table 1). This implies that effective approaches should be taken to enhance the production and use

Table 1 Total green vehicle sales and sale changes in Malaysia 2013–2015

	Sales			Changes in sales			
Car model	2015 (unit)	2014 (unit)	2013 (unit)	Δ unit (2014–2015)	Δ unit (2014–2015)	Δ unit (2013–2014)	Δ unit (2013–2014)
Audi	0	90	640	−90	−100	−550	−86
BMW	0	1	4	−1	−100	−3	−75
Honda	2	4,235	3,559	−4,233	−100	676	19
Lexus	7	100	299	−93	−93	−199	−67
Nissan	1,328	1,079	n/a	249	23	1,079	n/a
Toyota	1,996	502	2,301	1,494	298	−1,799	−78
Total	3,333	6,007	6,803	−2,674	−44.5	−796	−11.7

Source Malaysian Automotive Institute (2016)

of green vehicles since the demand for green vehicles is expected to rise rapidly due to its growing popularity [13]. Most of the people who lives in urban area are aware of the effects of motor vehicles on the environment. Environmental issues such as ozone depletion, acid rain, noise pollution and air contamination (Molina, 2004). Most of the users are knew the risks brought by the extensive use of motor vehicles to the environment. Unfortunately, none of those reasons stated earlier influence on people's behaviour despite having some evidence that indicates its negative effects on the environment [1, 6, 19]. It is not enough to change the behaviour of urban people as they are reluctant to utilize more sustainable ways of transportation such as walking or cycling.

Green vehicles are environment friendly products when we are talking of world-wide global warming. After so many dialogues and consideration, Malaysia has withdrawn the exception of the green vehicle import and excise duty starting year 2013. As a matter of fact, costs of green vehicles are much higher than the conventional ones. Green vehicles are just reasonable to those high-income purchaser groups, preventing other customers from acquiring them. However, the maintenance cost of green vehicles is relatively lower than the conventional vehicles unless the system of green vehicles breaks down and the warranty period has expired which makes it higher maintenance cost.

Based on the Transportation Policy in encouraging the use of public transportation, it proves to be relatively insignificant in overcoming the environmental issues [20]. Furthermore, rising gasoline prices also does not seem to reduce the number of private vehicles on the road [2]. On the account of petrol cost in Malaysia, the variance of the costs suggests no significant effects on the society, keeping in mind that the end goal is to diminish their dependence on the non-green vehicles [9]. However, the total vehicle sales in Malaysia continue to increase year by year, as depicted in Fig. 1. Consequently, the level of CO_2 emissions continues to increase. The fuel

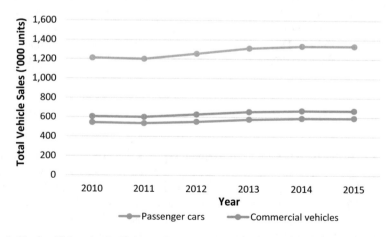

Fig. 1 Total vehicle sales in Malaysia between 2010 and 2015. *Source* Malaysia Automotive Institute (MAI) [11]

itself causes more damages, which include carbon discharges from the vehicles. The private vehicles on road remain high which suggests that Malaysians are still not aware about the negative impacts of the pollutants from the vehicles.

Therefore, this study investigates the attributes that are more of a priority among consumers towards green vehicles by using choice modelling (CM). Thus, respondents were asked to make a choice between two bundles of attributes with a related price. Consumers will derive satisfaction and preferences from the distinctive bundles of attributes provided in the questionnaires.

2 Literature Review

A green vehicle is viewed as the most inventive product in the automotive industry. It is valuable to lessen the effect of global warming because of discharges of CO_2 (Yon et al. [21]). Malaysian government has announced the use of green vehicles as a national priority and incorporated in the key research areas such as environmental concerns. It also forms innovative approaches to protect the environment [15].

The key procedure in the buyer's decision making is the combination procedure as per stated in Peter and Olson [12]. Bagozzi and Dholakia [4] depict purchase intention as a coordination of purchase goal processes under consideration with a brand or product qualities. Kotler et al. [10] expressed that vehicle products in the industry include complex purchasing behaviour as it includes high association of the consumers. Great publicizing image or reputation will encourage the high inclusion consumers to recall the vehicle products.

All in all, a high inclusion product will regularly utilize discerning promotion in advance. Such promotion upgrades the components of the product to be more unique than other products. The promotion activities could push the consumer purchase intention. In this matter, the buyers have a tendency by distinguishing the features, evaluating and perceived estimation of owning a product to assess the option decisions [5]. Stokes and Hallets [18] found that emotional experiences can be accomplished by the buyers and improve by the passionate interests depicted in a promotion activity.

2.1 Methodology

In this study, the investigation towards green vehicles was conducted by examining consumer preferences towards the ownership of green vehicles as far as emissions level, purchase price, vehicles range and source of energy are concerned. All attributes are crucial in order to identify the most preferred characteristics of green vehicles that would influence buyers' preferences in Malaysia. The first step in choice modelling (CM) is the identification of attributes and the level as found in previous studies related to the green vehicles. The 'status quo' term should also be included.

Table 2 Attributes and attributes levels

Attributes	Attributes levels
Emissions levels (CO_2)	(1) 100gm / km* (2) 90gm / km (3) 80gm / km
Vehicles range (km)	(1) 20 km/litre* (2) 25 km/litre (3) 30 km/litre
Source of energy	(1) Petrol* (2) Hybrid Vehicle (HV) (3) Electric Vehicle (EV)
Tax increase (price)	(1) Increase exise duty by 20%* (2) Increase exise duty by 10% (3) No increase in exise duty

Source Bakti and Normizan [3]
Note *Status quo or current situation of vehicles in Malaysia

The chosen attributes and levels of economic valuation of green vehicle technology preferences in Malaysia are presented in Table 2.

To represent the essential model behind the CM exhibited here, consider a respondent's choice for green vehicles. It assumes that the utility depends on the choices made from a set C namely a choice set. The choice set contains options which represent all the possible types of green vehicles. Alluding to Karousakis and Birol [8], the respondent is expected to have a utility function of the form:

$$U_{ij} = V\left(Z_{ij}, K_i\right) + e_i \tag{1}$$

Where:
U = utility (associated with any green vehicles criteria).
i = number of respondents.
j = types of green vehicles.
Z = attribute of green vehicles.
K = characteristics of the respondent (social, economic and attitudinal).
e_i = unobservable portion of utility (error component).

At the same time, characteristics are linear in the parameters and in the variable function. Furthermore, the error terms must be independently distributed with a Weilbull distribution. The marginal rate of substitution (MRS) is at which consumers are willing to trade-off between the attributes. The substitution rate can be estimated by dividing the β coefficient with another β coefficient (monetary) and then, multiplying it by –1. Hence, the equation yields as follows:

$$p_k^c = \frac{\frac{\partial v}{\partial xc,s}}{\frac{\partial v}{\partial Pc,s}} = \frac{-1\beta c, s}{\beta c, s = p}$$
$$= \frac{\beta\, attribute}{\beta\, monetary} \times -1 = \frac{\beta\, attribute}{\beta\, monetary} \tag{2}$$

This study incorporates a monetary attribute since it indicates the amount that the consumers are willing to pay to improve environmental quality.

3 Results and Discussion

The detailed distribution is shown in Table 3. The total number of respondents surveyed is 427; of which 215 are males and 212 are females. The age of the respondents ranges between 17 years old and above 60 years old. Of the total respondents, 58.1% are married while the rest are either single (40.3%), divorced (0.9%) or widowed (0.7%). Majority of the respondents are Malay (62.3%) and the rest are Chinese (18.7%), Indian (17.6%) and others (1.4%). The respondents also include a fair number of undergraduates (57.4%). Second in the list are diploma holders (23%), followed by postgraduates (15.7%). Lastly, 4.0% of the respondents have secondary level education. Their income level ranges between below RM2000 and above RM7000. From the list presented in Table 3, a significant number of the respondents are working in the private sector (64.6%), while 22.2% of them are working in the government sector and the rest are either self-employed (7.0%), working in semi-government sectors (4.4%) or having other working experience (1.6%).

The descriptive statistics which consist of the four attributes using CM method are presented in Table 4.

Table 4 presents the means and standard deviations for the attributes. The mean values range between 1.75 and 2.27 while the standard deviation values are between 0.48 and 0.73.

3.1 Conditional Logit (CL) Model

This part shows the CL incorporating levels with model specifications for improvement in consumer preferences towards green vehicles that consist of their attributes. The model is illustrated in the equation below:

$$U = \beta_1 X_1 + \beta_2 X_2 + \beta_3 X_3 + \beta_4 X_4 + \beta_5 X_5 + \beta_6 X_6 + \beta_7 X_7 + \varepsilon. \qquad (3)$$

Where: U = Utility.
$\beta_1 X_1$ = Lower level of emissions (CO22).
$\beta_2 X_2$ = Lowest level of emissions (CO23).
$\beta_3 X_3$ = Higher range of driving per litre (RNG2).
$\beta_4 X_4$ = Highest range of driving per litre (RNG3).
$\beta_5 X_5$ = Efficient source of energy (SRC2).
$\beta_6 X_6$ = Most efficient source of energy (SRC3).
$\beta_7 X_7$ = Additional cost of tax (PRICE).

Table 3 The socio-demographic findings

No	Demographic Variable	N	%
1	**Gender**		
	Male	215	50.4
	Female	212	49.6
2	**Age**		
	17–29 Years old	82	19.2
	30–39 Years old	129	30.2
	40–49 Years old	102	23.9
	50–59 Years old	88	20.6
	60 and Above	26	6.1
3	**Race**		
	Malay	266	62.3
	Chinese	80	18.7
	Indian	75	17.6
	Others	6	1.4
4	**Income**		
	RM0-RM2000	37	8.7
	RM2001-RM3000	29	6.8
	RM3001-RM4000	99	23.2
	RM4001-RM5000	102	23.9
	RM5001-RM6000	85	19.9
	RM6001-RM7000	42	9.8
	RM7001 and above	33	7.7
5	**Marital**		
	Single	172	40.3
	Married	248	58.1
	Divorced	4	0.9
	Widowed	3	0.7
6	**Education**		
	Secondary School	17	4.0
	Diploma	98	23.0
	Undergraduate	245	57.4
	Postgraduate	67	15.7
7	**Employment**		
	Government	95	22.2
	Semi-Government	19	4.4
	Private	276	64.6
	Self-Employed	30	7.0
	Others	7	1.6

Table 4 Descriptive statistics

Attribute	Attribute level	Freq (%)	Mean / Std.Dev	Min	Max
CO2	Emission levels				
	(1) 100gm/km*	2598 (50)	1.75 / 0.73	3	1
	(2) 90gm/km	1299 (25)			
	(3) 80gm/km	1299 (25)			
RNG	Vehicle Range				
	(1) 20 km/litre*	1299 (25)	2.64 / 0.48	1	3
	(2) 25 km/litre	1732 (33)			
	(3) 30 km/litre	2165 (42)			
SRC	Source energy				
	(1) Petrol*	2598 (50)	1.75 / 0.73	1	3
	(2) Hybrid	1299 (25)			
	(3) Electric	1299 (25)			
PRICE	Tax increase				
	(1) 20%*	1198 (23)	2.75 / 0.65	0	0.2
	(2) 10%	1398 (27)			
	(3) 0%	2600 (50)			

Note *Status quo or current situation of vehicles in Malaysia

The above parameters describe the importance of the attributes and their levels in determining consumers' preferences in selecting the best option for green vehicles. The consumers select the best option among the three different options offered. Respondents will choose the status quo if they do not intend to have any change in vehicle and they are very comfortable with the current situation. Table 5 presents the results of the CL simple model, which portrays consumers' options among the selected attributes in the study. The attributes also include the monetary attribute that is PRICE.

According to Table 5, all option attribute coefficients are highly significant at 5% level with an expected right sign. These results indicate that all attributes are inherent bias towards either the change or no-change options by respondents. PRICE is significant at 5% level, with the expected negative sign. It verifies that increase in additional cost in term of road tax towards the non-green vehicles has negative impacts on utility. The variable PRICE has a major impact on willingness to pay (WTP) and it affects selection of the best choice among other choices in the attribute level.

Based on the table, the level of independent variables are low as described in the model, since the PseudoR^2R^2 is 0.00918 (from a minimum value of 0 to a maximum of approximately 1). Gujariti (1999) points out that the larger the numbers of explanatory variables in the model, the higher will be the R^2. However, the statement should not be taken seriously since R^2 does not take into consideration "degree

Table 5 Result of conditional logit (CL) Model

Variables	Coefficient (β)	Std. Error	t-ratio	p-value	Marginal value (%)
CO22	0.8502***	0.0877	9.691	0	14.5333
CO23	0.5870***	0.2037	2.881	0.004	10.0341
RNG25	0.6106***	0.2325	2.625	0.0087	10.4376
RNG30	1.2005***	0.1824	6.579	0	20.5213
SRC2	0.7731***	0.1345	5.748	0	13.2154
SRC3	0.4294***	0.1313	3.268	0.0011	7.3402
PRICE	−0.0585***	0.0079	−7.399	0	
Summary Statistic					
No of observations		1732			
Log Likelihood		−1410.5			
Pseudo R^2		0.00918			
Adjusted Pseudo R^2		0.00717			

Notes ***Significant at 1%*, **Significant at 5%, *Significant at 10%

of freedom". The marginal rate of substitution (MRS) between other attributes and monetary attributes (PRICE) is regressed by using the WALD Test. MRS is the rate at which a consumer can give up some amount of good X (non-green vehicle) in exchange for good Y (green vehicle) while maintaining the same level of utility.

According to the results for marginal value, all attributes give positive sign which means positive relationship. CO22 and CO23 have marginal values of 14.5333 and 10.0341 respectively; this shows that the highest level of emissions will result in marginal value of 14.53 and 10.03%. The marginal value calculated for RNG25 and RNG30 indicates positive values of 10.4376 and 20.5213 respectively; which means that the increase in range of driving will positively give marginal value of 10.44 and 20.52%. Lastly, SRC2 and SRC3 have a positive relationship with the calculated marginal value of 13.2154 and 7.3402 respectively; which indicates that improvement in source of energy for vehicles have marginal value of 13.22% and 7.34%. As a conclusion, the highest marginal value is from RNG30, which means consumers will get more benefit in terms of range of driving (distance) from green vehicles compared to non-green vehicles by increasing rate of 20.52%.

3.2 Marginal Willingness to Pay (WTP)

The marginal willingness to pay (WTP) was calculated by computing the marginal rate of substitution, whereby this is the proportion of the coefficient of an attribute relative to the price coefficient. By using the marginal rate of substitution, WTP of respondents can be found according to their revealed preferences [17]. For example, in this study, one of the attributes was level of environmental observation, so by

Table 6 Marginal value for CL simple

Variables	Marginal value CL simple
CO22	14.5333
CO23	10.0341
RNG25	10.4376
RNG30	20.5213
SRC2	13.2154
SRC3	7.3402

dividing the β value of this attribute with the β value of the price, this would show the average willingness to pay of the respondents to increase the quality of environment at current level. The formula for marginal value is given as:

$$MV = \frac{-\beta\, attribute}{\beta\, Monetary\, value} \tag{4}$$

The Wald procedure contained in the LIMDEP 8.0 and NLOGIT 3.0 software was employed to estimate the WTP values of the attributes.

Result of Marginal Value (MV).

This study is significant since it estimates the monetary trade-off between all the attribute levels and price attribute. In this part, the differences of marginal values in all attribute levels are discussed. As such, the highlight is on the monetary trade-off between the two attributed levels and at the same time, it is believed that other attribute levels remain constant. The marginal values measured are in percentages (%). Table 6 shows the marginal value for CL simple models.

Based on Table 6, the marginal value calculated for CO22 from the CL simple model is 14.53, which indicates that each unit increase in level of emission has a marginal value of RM14.53. The positive sign indicates that the utility has been increased or respondents are willing to pay 14.53% increase in price to reduce the level of emission from the current level. The marginal value for CO23 is 10.03 in the simple model. Respondents are willing to pay for 10.03% increase in price for very good level of emission. Based on a range of driving, the marginal value for RNG25 is 10.44 in the simple model. The positive sign indicates that utility has increased. This means that each unit increase in the range of driving has a marginal value of 10.44% in conservation price for the simple model. The marginal value for RNG30 is 20.52 in the simple model. For the simple CL model, respondents are willing to pay an increase of 20.52% in conservation price for highest range of driving offered by green vehicles.

On the other hand, for source of energy attributes, the marginal value for SRC2 in the simple CL model is 13.22. This means that each unit increase in source of energy has a marginal value of 13.22% of conservation price; or respondents are willing to pay that amount of increase in conservation price to increase the quality of the environment from the current level. The variable SRC3 shows a positive marginal value of 7.34 in the simple model. The highest marginal value or willingness to pay (WTP) of the respondents for the CL simple from the range of driving, which is RNG30 (20.52%).

4 Conclusion

The analysis of green vehicle preferences using the choice modelling (CM) method showed that the highest marginal value was 20.5% from range of driving. It means that most of the respondents prefer the green vehicles that could give more distance compared to other characteristics. All variables showed significant values as well as positive relationships. This study also demonstrates CM approaches to reveal the attributes that influence choices. In other words, the CM experiments offer a sequence of choices that enables respondents to choose based on the desirable attributes. This means that the greater the satisfaction or utility associated with a particular option, the greater the probability that it would be chosen by the respondents. Besides that, the CM can assess benefits transfer when environmental goods have measurable attributes that can be estimated [7]. These valuations are significant since they offer information and knowledge regarding green vehicle attributes and can be a good source for planning any developments of services in the future.

The outcome of the CM methods suggests that the green vehicle attributes are significant, indicating that respondents made their choices between options with reference to the attribute level. 'Range of Driving' or distance of the vehicle per litre attribute becomes the most significant as people are willing to pay a fine, producing the highest marginal value compared with other attributes.

Based on the past literature, even though there are several studies that employed stated preference methods to estimate the economic value of green vehicles towards the environment, little attention was given to the CM method for it. The CE presented in this study provides four attributes, including level of emissions, range of driving, source of energy and additional tax (price). The CE calculates respondents' green vehicle preferences and provides benefit transfer potential based on attributes and levels. According to the findings, people can contribute more on the "Range" attribute because it provides the highest marginal benefit. It is clearly portrayed that people wish to improve the environment condition by having a green vehicle that can give the highest satisfaction particularly in terms of range or distance of driving.

Environmental challenges have put so much pressure on governments to find ways to reduce environmental damage while lessening the harm to economic growth. There are many important advantages of environmental taxes, such as environmental effectiveness and transparency. The government should be thinking about using more resources to reduce air pollution by carrying out the new tax system such as the mandatory payment method on road tax price for non-green vehicles. The funding from this payment will go to environmental research and development programs, the recovery fund from the pollution, the protection fund to address the various environmental problems.

References

1. Anable, J.: 'Complacent car addicts' or 'aspiring environmentalists'? Identifying travel behaviour segments using attitude theory. Transp. Policy **12**(1), 65–78 (2005)
2. Ariffin, R.N.R., Zahari, R.K.: The challenges of implementing urban transport policy in the Klang Valley, Malaysia. Proc. Environ. Sci. **17**, 469–477 (2013)
3. Bakti, H.B., Normizan, B.: The effect of knowledge types on hybrid cars preferences. In: Proceeding of the International conference on social and economic development (ICSED 2016), pp 270–219 (20)
4. Bagozzi, R P., Dholakia, U.: Goal setting and goal striving in consumer behavior. J. Market. **63**(4_suppl1), 19–32 (1999)
5. Belch, G.E., Belch, M.A.: Advertising and Promotion, An Integrated Marketing Communication Perspective, 9th ed, pp. 290–294. McGraw-Hill. Singapore (2012)
6. Hagman, O.: Mobilizing meanings of mobility: car users' constructions of the goods and bads of car use. Transp. Res. Part D **8**(1), 1–9 (2003)
7. Hanley, N., Wright, R.E., Adammowicz, W.: Using choice experiments to value the environment: design issues, current experiment and future prospects. Environ. Resour. Econ. **11**(3–4), 413–428. https://doi.org/10.1023/A:1008287310583
8. Karousakis, K., Birol, E.: Investigating household preferences for kerbside recycling services in London: a choice experiment approach. J. Environ. Manage. **88**(4), 1099–1108 (2008)
9. Kingham, S., Dickinson, J., Copsey, S.: Travelling to work: will people move out of their cars. Transp. Policy **8**(2):151–160 (2001). https://doi.org/10.1016/S0967-070X(01)00005-1
10. Kotler, P., Amstrong, G., Swee, H.A., Siew, M., Chin, T.T., David, K.T.: Principles of Marketing, An Asian Perspective. Prentice Hall, Pearson Education South Asia, London, United Kingsom (2006)
11. Malaysia Automobile Industry: MAI website data on sales of green vehicles in Malaysia for year 2012–2015 (2016). Retrieved 1 April 2018 from https://www.miti.gov.my/miti/resources/Industry4Point0/MAI_New_Horizon_of_Data_Driven_Economy.pdf
12. Peter, J.P., Olson, J.C.: Consumer Behaviour and Marketing Strategy, 6th edn. McGraw-Hill, London (2002)
13. Rosli, M., Kari, F.: Malaysia's national automotive policy and the performance of proton's foreign and local vendors. Asia Pac. Bus. Rev. **14**(1), 103–118 (2014)
14. Sampaio, M.R., Rosa, L.P., Márcio de Almeida, D.: Ethanol–electric propulsion as a sustainable technological alternative for urban buses in Brazil. Renew. Sustain. Energy Rev. **11**(7), 1514–1529 (2007). https://doi.org/10.1016/j.rser.2005.11.007
15. Taghizadeh, S.K., Jayaraman, K., Ismail, I., Rahman, S.A.: A study of service innovation management in the Malaysian telecommunications industry. Glob. Bus. Organ. Excell. **34**(1), 67–77 (2014)
16. Taghizadeh, S.K., Jayaraman, K., Ismail, I., Ahmad, N.H., Rahman, S.A.: Innovation Value Chain as Antecedent of Service Innovation Management Practices: Experience from Malaysian Telecommunication Sector. Penang, Malaysia (2014). Retrieved from: https://www.researchgate.net/profile/Syed_Abidur_Rahman/publication/283047920
17. Siebert, F., Hildebrandt, P.: Vibrational Spectroscopy in Life Science. John Wiley & Sons, New Jersey, United State (2008)
18. Stokes, G., Hallet, S.: The roles of advertising and the car. Transp. Rev. **12**(2), 171–183 (1992)
19. Tertoolen, G., Van Kreveld, D., Verstraten, B.: Psychological resistance against attempts to reduce private car use. Transp. Res. Part A **32**(3), 171–181 (1998)

20. Timilsina, G.R., Dulal, H.B.: Urban road transportation externalities: costs and choice of policy instruments. World Bank Res. Obs. **1**, 162-191 (2011)
21. Yong, H.H., Khan, N., Abd, M.M.: The determinants of hybrid car adoption: Malaysia perspective. Aust. J. Basic Appl. Sci. **7**(8), 347–454 (2012)

Relationship Between Corruption, Governance, and Economic Growth in ASEAN

Toh Kim Yuan and Suryati Ishak

Abstract Corruption has become an issue in ASEAN and it bring a negative impact on ASEAN economic growth. On the other hand, quality of governance closed related to corruption. Poor governance will increase corruption. Furthermore, good governance is important to contribute to a stable nation and economic growth. So, this study aims to investigate the relationship between corruption, governance and economic growth in five ASEAN which include (Singapore, Malaysia, Thailand, Indonesia and Philippines). The purpose of this study is to examine the long run relationship between economic growth, corruption and governance in ASEAN 5. This study used static panel data analysis to analyse the relationship. The results show that corruption related to the gross domestic product per capita. The coefficient of corruption is statistically significant mean according the economic theory that corruption brings negative effect to economic growth.

Keywords Corruption · Governance · Panel data · ASEAN

Association of Southeast Asia Nations (ASEAN) was established on 8 August 1967. Initially, Indonesia, Malaysia, the Philippines, Singapore, and Thailand, decided to form ASEAN and signed the document in Bangkok, Thailand. Five representatives from the Foreign Ministry of each country signed the ASEAN Declaration. Economic growth is an important goal in ASEAN. The purpose of ASEAN is to focus on the ASEAN members' economic, social, and cultural development. ASEAN also promotes political stabilisation among its members and increases its power. Additionally, ASEAN also focuses on educational quality by assisting each other in terms of training and facilities.

According to Jennifer Schoeberlein [8], the ASEAN region countries are among the fastest-growing economies globally. A significant increase in foreign direct investment and regional integration can be seen in recent years. However, despite

T. K. Yuan · S. Ishak (✉)
School of Business and Economics, Faculty of Economics and Management, Universiti Putra Malaysia, Serdang Selangor, Malaysia
e-mail: suryatiis@upm.edu.my

© Institute of Technology PETRONAS Sdn Bhd 2022
S. A. Abdul Karim (eds.), *Shifting Economic, Financial and Banking Paradigm*,
Studies in Systems, Decision and Control 382,
https://doi.org/10.1007/978-3-030-79610-5_8

economic growth, sustainable development in the region is hampered by severe governance shortcomings, most notably in autocratic governments, low accountability levels, and highly politicised public sectors.

1 Introduction

Corruption occurs as a result of institutional weaknesses in a country. It happens when the public sector abuses its authority for private advantage, which eventually will impact economic growth. Thus, corruption can lead to a decline in economic development, resulting in poor economic growth (Freckleton et al. [4].

According to the World Bank, corruption is an abuse of public office for personal benefits. This definition of corruption does not discharge and acquit the private sector from being corrupt, especially when it comes to awarding contracts, procurement, or hiring in large companies, particularly multinational companies.

According to the Transparency International report [10], Singapore was the top least corruption problem among five ASEAN countries with Corruption Perceptions Index (CPI) scores between 84 and 94 from 1996 to 2017. Malaysia CPI score had an inconsistent trend with scores between 47 and 54 from 1996 to 2017. Based on the report, Malaysia was recognised as the second least corrupted nation among ASEAN countries. Meanwhile, Thailand, the Philippines, and Indonesia experienced an uncertain trend in the CPI score.

1.1 *Governance*

Many researchers have different concepts of governance but still do not have a precise definition of governance. Governance plays a significant role in a country to set and execute rules and regulations to restrict citizens' actions to control their ethical and moral behaviour. It also plays a role in managing a country from economic, political and social perspectives.

According to the World Bank, governance consists of control of corruption, political stability, the voice of accountability, the rule of law, regulatory quality, and government effectiveness. Based on the UNDP discussion paper [11], good governance outcomes could be peaceful, stable, and resilient societies in which services are delivered and reflect communities' needs, including the most vulnerable and marginalised voices.

In the meantime, the subject of governance and corruption issues have been observed in all ASEAN region and globally. Governance performance is closely related to corruption issues. Governance performance can be measured by governance indicators: the rule of law, political stability, control of corruption, government effectiveness, voice and accountability, and regulator quality. One of the governance failures is the issue of corruption.

Table 1 Corruption perception index in ASEAN (1996-2017)

Year	Indonesia	Malaysia	Phil	Singapore	Thailand
1996	2.65	5.32	2.69	8.8	3.33
1997	2.72	5.01	3.05	8.66	3.06
1998	2	5.3	3.3	9.1	3
1999	1.7	5.1	3.6	9.1	3.2
2000	1.7	4.8	2.8	9.1	3.2
2001	2.9		2.9	9.2	3.2
2002	1.9		2.6	9.3	3.2
2003			2.5	9.4	3.3
2004			2.6	9.3	3.6
2005			2.5	9.4	3.8
2006			2.5	9.4	3.6
2007			2.5	9.3	3.8
2008			2.3	9.2	3.6
2009			2.4	9.2	3.3
2010			2.4	9.3	3.5
2011			2.6	9.2	3.4
2012			3.4	8.7	3.7
2013			3.6	8.6	3.5
2014			3.8	8.4	3.8
2015			3.5	8.5	3.8
2016			3.5	8.4	3.5
2017			3.4	8.4	3.7

Source Transparency International

Government institution needs to protect its citizens' property right. Good governance can help to achieve economic growth rapidly by using sound capital and labour.

Table 1 shows the Corruption Perceptions Index (CPI) based on the perception levels of public sector corruption by experts and business people in ASEAN countries. Transparency International defines the CPI scale ranging from 100 as very clean to 0 as highly corrupt.

2 Problem Statement

ASEAN economic growth has been slower than in developed countries. Several factors affect economic growth; however, internal factors influencing a nation's economic growth are quality of governance, corruption, and social problems. Some of the ASEAN countries face poor governance problem and higher corruption that can negatively affect economic growth.

Economic growth in Malaysia, Thailand, Indonesia, and the Philippines is slower than in Singapore due to the political crisis from 1996 to 2017. Malaysia, Thailand, Indonesia and the Philippines perform worse in governance indicators, with a low CPI score (high corruption). On the other hand, Singapore performs well with a high CPI score (low corruption) in governance indicators and sound economic growth performance.

Corruption will increase when the quality of governance declines, affecting economic growth. For example, from 2013 to 2014, there was political instability in Thailand due to its Prime Minister, Yingluck, involved in corruption. The Prime Minister involvement in government rice subsidy corruption led to dropped CPI but increased economic growth. Thus, this study investigates the relationship between the quality of governance and corruption on economic growth.

3   Objective of the Study

The purpose of this study is to examine the long-term relationship between economic growth, corruption, and governance in ASEAN 5.

4   Empirical Findings

Corruption and governance on economic growth

Many studies indicated that corruption negatively affects economic growth. Several empirical studies claimed a negative relationship between corruption and governance on economic growth, influenced by the governance indicator.

According to Brewer, and Walker [2], There are six dimensions of governance are included in the set of indicators: Voice and accountability, Political stability and absence of violence, Government effectiveness, Rule of law, Regulatory quality, and Control of corruption. It is because these three indicators are more related to public management. Based on the result, non-Asia countries showed that democratic government is not correlated with corruption. America and Europe performed well in governance indicators; however, Asia was ranked last in voice and accountability, fifth in control of corruption, and midpoint in government effectiveness. The research found that good governance could control corruption and positively impact economic growth.

Tareq and Ahmed [9] focused on the impact of governance on corruption and economic growth in 30 developing countries. The results of the six governance indicators showed a significant relationship with economic growth. The quality of governance is important to economic growth, while good control of corruption can mitigate the negative effect of corruption and boost economic growth. Government effectiveness also promotes the business climate and encourages more investment in developing countries.

Dagostino, Dunne, and Pieroni [3] found that corruption causes an inefficient allocation of government spending in the military sector in their research on government spending, corruption, and economic growth. The authors used the panel data because the research included 106 countries and many years. They argued that government spending is resulted from improving economic growth. The findings showed that the government spending of 106 countries focuses more on the military and that more government spending would boost economic growth. However, corruption caused government spending to be less important on economic growth because most government spending was due to private benefits, not for developing the country.

5 Endogenous Growth Theory

The endogenous growth model was developed by economist Romer [7]. The endogenous growth theory describes a direct result of an internal process that will influence economic growth. Some of the elements, such as human capital, innovation, and knowledge used to measure long-term growth, depend on policy measures.

The endogenous growth model is related to corruption, governance and economic growth. This theory indicates that low corruption will increase the investment in a country's human capital, helping develop the technology and increase production and economic growth for a nation. Other economists mentioned that investment in human capital would help to promote innovation and increase productivity. The public and private sectors provide the human capital incentive to be more creative in contributing to economic growth.

The endogenous growth model uses the AK model of a Cobb-Douglas production function to describe long-term growth:

$$Y = AK^a L^{1-a}$$

where,

Y is the total output in the economy, A is technology, K is capital, and L is labour.

The endogenous growth model explains that internal factors affect economic growth rather than external factors in the neoclassical model. The endogenous growth theory focuses on long-term economic growth. It includes physical capital and education.

Barro's model adds another element to the AK model, which is government spending:

$$Y = AGK^a L^{1-a}$$

where,

Y is the total output in the economy, G is government spending, A is total productivity, K is capital, and L is labour.

6 Model Specification

To observe the relationship between corruption, governance, and economy in ASEAN 5, the researcher followed the model used by Omoteso and Ishola Mobolaji [6].

This study aims to examine the impact of corruption and governance on the economy of ASEAN countries. The general form can examine the role of the specified independent variables on ASEAN 5's economic growth.

The functional form is formulated as follows:

$$Y = f\ (CPI, EI, POS) \tag{1}$$

where,

Y is the GDP per capita used as a proxy for economic growth and is assumed to be affected by the level of corruption. CPI refers to the corruption perception index, EI refers to the education index, and POS refers to political stability.

The GDP per capita is transformed into a log form:

$$LnRGDP_{it} = \beta 0_i + \beta 1 CPI_{it} + \beta 2 EI_{it} + \beta 3 PS_{it} + \varepsilon_{it} \tag{2}$$

where,

the equation Ln describes the logarithm form of real GDP per capita, and $\beta 0$ represents the intercept, differing across countries in the fixed effects model. $\beta 1, \beta 2$, and $\beta 3$ are slope coefficients of corruption perception index, education index, and political stability. These coefficients are taken as fixed/constant for each cross-section in the fixed effects model, whereas ε_{it} is the error term for each cross-section entity.

6.1 Variables Description and Measurement

Gross Domestic Product Per Capita (GDP)

This study's dependent variable is GDP per capita (constant 2010 US$), taken from World Development Indicators (WDI).

$$GDP\ per\ capita = \frac{GDP}{population}$$

GDP per capita is GDP divided by population. The GDP components include the total amount of gross value, item taxes and deduct subsidies. However, it does not include the depreciation of items. All data are calculated in constant local currency.

Governance Indicators

The World Governance Indicators (WGI) compile and summarise information from 214 countries existing data sources around the world on the quality of different governance aspects. The WGI includes six aggregate governance indicators. This study used only two out of the six measures of governance indicators, as proposed by the World Bank (Kaufman et al. [5], including political stability and control of corruption.

Corruption Perception Index (CPI)

One independent variable is the Corruption Perception Index (CPI). Transparency International provides the CPI. The CPI ranges from 0 to 10, in which 0 and 10 scores indicate the highest and lowest degree of corruption, respectively.

Political Stability

Political stability refers to the conflict or competition between the government and the opposition. The estimates of political stability are taken without log because these values vary between -2.5 (poor quality of governance) and $+2.5$ (good quality governance).

Education Index

The education index is one of the components of Human Development reports from United Nations Development Programme. The average and expected years of schooling equal to the education index.

Education Index $=$ years of schooling $+$ expected years of schooling index2

One of the components to determine economic development and quality of life is education. The adult literacy rate and the combined primary, secondary, and tertiary gross enrolment ratio are used to determine the education index. The mean year of schooling and expected years of schooling is used to calculate the education index.

The education Index is one of the independent variables because this study also focuses on the relationship between higher education and corruption on economic growth. Higher educated people are less involved in corruption or more involved in corruption.

6.2 Data Sources

The current study used panel data for ASEAN 5 member countries: Singapore, Thailand, Indonesia, Malaysia, and the Philippines. The period used in the current study

Table 2 Summary of data set used

Variables	Descriptions	Expected Signs	Sources
GDP per capita (constant in 2010)	Real GDP per person in a country		World Bank
Corruption perception index	The corruption perception index published by Transparency International, which scores the levels of public sector corruption by their countries.	Positive	Transparency International
Political stability	The conflict or competition between the government and the opposition	Positive	World Governance Indicators (WGI)
Education index	The adult literacy rate and the combined primary, secondary, and tertiary gross enrolment ratio	Positive	Human Development Reports

was 1996–2017. The relevant data for variables were taken from World Development Indicators (WDI) and World Governance Indicators (WGI).

GDP per capita (constant in 2010) was taken from the WDI. The WGI compiled and summarised information from five existing data sources around ASEAN on the quality of various governance aspects. The WGI included six aggregate governance indicators. This study only used political stability. These governance variables were taken without log because their values varied between –2.5 (poor quality of governance) and +2.5 (good quality of governance) (Table 2).

7 Panel Data Analysis

This study used the panel data method, combining time-series and cross-sectional data. The study chose the panel data technique to allow for differences in unobservable individual country effects. A panel study's ability to control individual heterogeneity and state and time-invariant variables makes it a superior technique for a time series or cross-sectional study [1]. In this study, the fixed effect (FE), random effect (RE), and Hausman tests (based on the difference between FE and RE estimators) were conducted. The Hausman test was employed to select the best model between the fixed effects model and the random effects model.

7.1 Discussion of Empirical Results

Descriptive Statistics

Table 3 shows the descriptive statistics for gross domestic product per capita, corruption perception index, education index in ASEAN 5 from 1996 to 2017. The Table 3 shows the total number of observations for all variables along with the mean, standard deviation, minimum and maximum values. The mean score of gross domestic product per capita in ASEAN 5 is 8.717367. The total average score of corruption is 4.601727, and the education index is 0.7214182. Meanwhile, the overall mean score of political stability is -0.2959091.

Fixed Effects Model

Table 4 depicts the fixed effect model panel data results in the correlation between corruption, governance, and economic growth in ASEAN 5. The results show that the corruption perception index is positively linked to the gross domestic product per capita. The coefficient of corruption is statistically significant, meaning that corruption negatively affects economic growth, according to economic theory. The corruption perception index indicates that a one-unit increase in corruption perception leads to a 0.56% increase in gross domestic product per capita in ASEAN 5. Higher corruption perception index means low corruption will increase economic growth. Next, the education index's coefficient has a significant and positive relationship with the gross domestic product per capita. The coefficient shows that a one-unit increase in the education index will increase economic growth by 4.93%. Political stability also indicates a significant and positive relationship with the gross domestic product

Table 3 Variables descriptive statistics summary

Variables	Obs	Mean	Maximum	Minimum	S.D.
LGDPPC	110	8.717367	10.94625	7.352224	1.09932
CPI	110	4.601727	9.4	1.7	2.380332
EI	110	0.7214182	0.932	0.577	0.0916824
PS	110	−0.2959091	1.59	−2.09	1.035755

Table 4 Results of fixed effects model dependent variable: LNLDPPPC

Variables	Coefficient	Standard-error	t-value	p-value
CPI	0.0561472	0.135078	4.16	0.00
EI	4.927653	0.1420543	34.69	0.00
PS	0.0260166	0.0113543	2.29	0.024
C	4.911793	1.1049414	46.05	0.00
R^2	0.95			
F-statistic		466.37(0.00)		

Table 5 Results of random effects model dependent variable: LNLDPPPC

Variables	Coefficient	Standard-error	t-value	p-value
CPI	0.0671035	0.0140626	4.77	0.00
EI	4.952611	0.1509635	32.61	0.00
PS	0.0274175	0.0120717	2.27	0.023
C	4.942919	0.1657204	26.7	0.00
R^2	0.95			
F-statistic		1290/32(0.00)		

per capita; the coefficient shows that a one-unit increase in political stability will increase economic growth by 0.026. The fixed effects model shows that R-square is 0.953, meaning 95% variation in GDP per capita is explained by the explanatory variables. The F-statistic is a high significance and good fit of this model.

Random Effects Model

Table 5 shows all variables of the random effect model (corruption perception index, education index, and political stability) are significant and positively impact GDP per capita in ASEAN 5. The corruption perception index, education index, and political stability's coefficients are statistically significant; a one per cent increase will increase economic growth by 0.67, 4.95, and 0.027, respectively. The R-square value is 0.956, meaning 95.6 per cent variation in economic growth is due to explanatory variables (corruption perception index, education index, and political stability). The F-statistic shows a significance and good fit of all the coefficients in this model.

Hausman Test

Hausman test was employed to test model misspecification. Hausman test helped observe whether the fixed effects model or random effects model is more appropriate for these panel data. Using the Hausman test, the p-value was less than 0.05; thus, the null hypothesis that the random effects model is more appropriate is rejected. The results indicated that the fixed effects model is better than the random effects model.

Overall, the outcome showed that the corruption perception index has a positive relationship with economic growth. Higher corruption perception index means less corruption occurs, and economic growth becomes better off. The education index and political stability also showed a positive relationship with economic growth. Some policy implications, limitations, and recommendations will be described in the last part of this chapter.

8 Summary of the Findings

In short, the relationship between corruption and economic growth is negative, while between governance and economic growth is positive. At the beginning of this research, the Breusch and Pagan Lagrangian Multiplier Test showed that the random effect exists. Then, the Hausman test showed that the fixed effects model is more appropriate for these panel data. Based on this test, the dependent or independent variables were found to be correlated.

Subsequently, the research proceeded with various diagnostic tests to check the multicollinearity, heteroscedasticity, and serial correlation issues. It was found that a serial correlation issue exists in the panel data; a third method, Daniel Hoechle, was employed to rectify the issue.

The Least Squares Dummy Variable (LSDV) estimated the period of dynamic panel data models more than the countries. The LSDV result showed that the country's effect existed while the effect of time did not exist.

These findings were according to prior expectations. All three variables, the corruption perception index, education index, and political stability, showed a positive and significant influence on economic growth for ASEAN 5.

8.1 Policy Implications

The policy implications for this study are straightforward. Corruption and quality of governance will have an impact on economic growth. Thus, improving the quality of governance and decreasing the corruption problem will improve economic growth.

First, the findings showed that corruption negatively impacts economic growth. The government should have a legitimate system to control corruption. Some bodies need to be established to control the corruption problem. Anti-corruption laws must be effective to combat corruption. Besides, young generations should be taught to fight corruption.

It is crucial to ensure political stability through governance efforts to control society's conflicts and violence. The government should play an important role to provide the best quality services to civilians. Policymakers should be strict in controlling political instability and corruption.

8.2 Limitations of the Study and Recommendations for Future Studies

First of all, there is a limitation in the secondary data available on the World Bank database used in this study. This study only used data from 1996 to 2017 since the World Bank database could only provide data for that period for ASEAN 5. Previous studies also provided data beginning in 1996. Future researchers are recommended to source data from other databases instead of only focusing on the World Bank database.

Next, the scope of this study is limited to five ASEAN countries and two governance indicators. Future studies can expand to other ASEAN countries and other governance indicators since no previous studies researched ASEAN countries.

Finally, this study only includes three independent variables: the quality of governance (political stability), corruption perception index, and education index. Future research can consist of more variables to examine the relationship between the variables and economic growth and generate more significant results.

References

1. Baltagi, B.H.: Econometric Analysis of Panel Data. John Wiley and Sons, New York, NY (1995)
2. Brewer, G.A., Walker, R.M.: Accountability, corruption and government effectiveness in asia: an exploration of world bank governance indicator 8(2) (2007)
3. D' Agostino, G., Dunne, J.P., Pieroni, L.: Government spending, corruption and economic growth. World Develop. 84, 190–205 (2016)
4. Freckleton, M., Wright, A., Craigwell, R.: Economic growth, foreign direct investment and corruption in developed and developing countries. J. Econ. Stud. 39(6):639–652 (2012)
5. Kaufmann, D., Kraay, A., Mastruzzi, M.: Governance matters VII: aggregate and individual governance indicators, 1996–2007. World Bank policy research 2008
6. Omoteso, K., Ishola Mobolaji, H.: Corruption, governance and economic growth in Sub-Saharan Africa: a need for the prioritisation of reform policies. Soc. Responsib. J. 10(2), 316–330 (2014)
7. Romer, P.M.: Increasing returns and long-run growth. J. Political Econ. 94(5), 1002–37 (1986)
8. Schoeberlein, J.: Corruption in ASEAN Regional trends from the 2020 Global Corruption Barometer and country spotlights (2020). Retrieved from https://knowledgehub.transparency. org/assets/uploads/kproducts/Corruption-in-ASEAN-2020_GCB-launch.pdf
9. Tareq, B.A., Ahmed, Z.: Governance and economic performance in developing countries: an empirical study. J. Econ. Stud. Res. 2013, 1–13 (2013)
10. Transparency International report (2018). Retrieved from https://www.transparency.org/en/cpi/ 2018/index/dnk
11. UNDP 2014-Governance for Sustainable Development Integrating Governance in the Post-2015 Development Framework. Retrieved from https://www.file:///C:/Users/user/Downloads/ Discussion-paper-Governance-for-Sustainable-Development%20(2).pdf

EVFTA—A Review of Opportunities and Challenges for Vietnam

Nguyen Thi Lieu Trang

Abstract EU—Vietnam Free Trade Agreement (EVFTA) was signed on June 30, 2019 and approved by the European Parliament (EP) on February 02, 2020 and the National Assembly of Vietnam has officially ratified the EVFTA on June 08, 2020, marking a significant milestone for the trade relations between Vietnam and the European Union. EVFTA is a comprehensive new generation agreement and the first EU's FTA with a developing country like Vietnam. This paper provides a review of new opportunities, challenges for and tasks of Vietnam once this agreement comes into enforcement.

Keywords EVFTA · EU · Vietnam · FTA

1 Background of EU—Vietnam Relationship

The EU is the region making up a large proportion of the trade relation between Vietnam and Europe. Vietnam—EU trade relation has developed very rapidly and effectively. According to the General Department of Vietnam Customs, from 2000 to 2017, the trade turnover between Vietnam and the EU increased by more than 13.7 times, from US\$4.1 billion in 2000 to \$56.45 billion in 2019; in which Vietnam's exports to the EU increased by 14.8 times (from \$2.8 to \$41.54 billion), and imports into Vietnam increased by 11.4 times (\$1.3–\$14.90 billion).

Especially in 2019, the import–export turnover between Vietnam and the EU reached over \$56.45 billion, increased by 1.11% to the same period in 2018, of which exports reached over \$41.54 billion (decreased by 0.81%), and imports reached \$14.90 billion (increased by 6.84%). The markets with the export value of more than \$1 billion in 2019 are the Netherlands (\$6.88 billion, decreased by 2.89% compared to 2018), Germany (\$6.56 billion, decreased by 4.63%), England (\$5.76 billion, decreased by 0.38%), France (\$3.76 billion, decreased by 0.01%), Italy (\$3.44 billion, increased by 18.46%), Austria (\$3.27 billion, decreased by 19.93%), Spain (\$2.72

N. T. L. Trang (✉)
Department of International Business, FPT University, Hanoi, Vietnam
e-mail: trangntl7@fe.edu.vn

© Institute of Technology PETRONAS Sdn Bhd 2022
S. A. Abdul Karim (eds.), *Shifting Economic, Financial and Banking Paradigm*,
Studies in Systems, Decision and Control 382,
https://doi.org/10.1007/978-3-030-79610-5_9

billion, increased by 3.38%), Belgium ($2.55 billion, increased by 5.83%), Poland ($1.50 billion, increased by 12.42%) and Sweden ($1.18 billion, increased by 2.39%).

For many years, the European Union (EU) has always been a major importer, with purchasing power ranked second in the world and the key market for Vietnamese exports. Total two-way trade turnover between Vietnam and EU in the year of commercial goods and taxes. During 10 years since the official take effect, EVFTA will eliminate nearly 99% of tariff lines and trade barriers between Vietnam and EU. This is a huge advantage compared to Vietnamese competitors such as Thailand, China, when they reached 56.5 billion USD in 2019, accounting for 10.9% of the total import–export turnover of the country, nearly 3 times higher than the rate of 17.6 billion USD in 2010. In which, export turnover reached 41.5 billion USD (accounting for 15.7%), import turnover reached USD 15 billion (accounting for 5.9%). Particularly in the first 11 months of 2020, the total two-way turnover between Vietnam and the EU reached 45.1 billion USD, accounting for 9.2% of the total import and export turnover of the country, of which export turnover reached 31.9 billion USD. USD (accounting for 12.5%) and import turnover reached 13.2 billion USD (accounting for 5.6%). See Table 1.

Recently, Vietnam's primary destinations for exports in the EU market have remained in traditional markets such as the Netherlands, Germany, the UK, France, Italy, Spain, Belgium, and Poland. For the Austrian market, export turnover in this market is mainly accounted for by the exports of mobile phones.

About export.

In 2019, Vietnam's exports to the EU reached US$41.54 billion, decreased by 0.81% compared to 2018. The main export items of Vietnam to the EU are phones and components (making up $12.21 billion, decreased by 7.23%), footwear ($5.03 billion, increased by 7.51%), computers, electronic products and components ($4.66 billion, decreased by 8.13%), textiles ($4.26 billion, increased by 3.90%), machinery, equipment and other spare parts ($2.51 billion, increased by 21.63%), fishery products ($1.25 billion, decreased by 13.07%), and coffee ($1.16 billion, decreased by 14.91%). The items with the highest growth in 2019 are plastic materials (reaching

Table 1 Import and export turnover Vietnam—EU statistics

Year	Import		Export		Import–export	
	Value	Increase (%)	Value	Increase (%)	Value	Increase (%)
2015	30.940,1	10,77	10.433,9	17,16	41.374,0	12,31
2016	34.007,1	9,92	11.063,5	6,03	45.070,7	8,93
2017	38.336,9	12,75	12.097,6	8,57	50.434,5	11,72
2018	41.885,5	9,42	13.892,3	13,95	55.777,8	10,59
2019	41.546,6	−0,81	14.906,3	7,30	56.452,9	1,21

Unit USD million
Source The general department of Vietnam customs

Table 2 Some primary export goods from Vietnam to EU

	Goods	2017	2018	2019	2019/2018 (%)
01	Footwear	4.612,3	4.677,8	*5.029,4*	**+7,51**
02	Textiles	3.733,3	4.101,7	*4.261,9*	**+3,90**
03	Fisheries	1.422,1	1.435,2	*1.247,6*	**−13,07**
04	Coffee	1.365,4	1.360,5	*1.157,7*	**−14,91**
05	Wood products	751,4	779,1	*846,6*	**+8,65**
06	Computers	4.097,5	5.072,9	*4.660,4*	**−8,13**
07	Phones	11.778,0	13.161,4	*12.209,2*	**−7,23**
08	Bags, wallets, suitcases, hats and umbrellas	879,5	929,8	*965,6*	**+3,85**
09	Steel products	399,8	568,8	*551,4*	**−3,06**
10	Means of transportation parts	705	671,6	*814,3*	**+21,24**
11	Cashew	944,4	105,4	*102,6*	**−2,66**
12	Machinery	1.688,4	2.063,8	*2.510,3*	**+21,63**

Unit USD million
Source The general department of Vietnam customs

$19.13 million, increased by 235.42%), paper and paper products ($13.94 million, increased by 175.56%), cameras, cine cameras and accessories ($30.70 million, increased by 139.83%), tea ($8.20 million, increased by 132.98%) and wires and cables ($31.10 million, increased by 139.83%). It is noteworthy that some export items decreased, for instance, iron and steel ($238.28 million, decreased by 33.98%), chemicals ($38.35 million, decreased by 16.83%), and rubber ($113.77 million, decreased by 11.37%), sea foods ($1.25 billion, decreased by 13.07%) and coffee ($1.16 billion, decreased by 14.91%). Table 2 summarizes all the data.

About import

In 2019, the import of goods from the EU reached US$14.90 billion, increased by 6.84% compared to 2018 (see Table 3). The main import items of Vietnam from the EU are machinery, equipment, tools and other spare parts (reaching $3.91 billion, decreased by 3.92%), computers, electronic products and components ($2.51 billion, increased by 36.40%), pharmaceutical products ($1.63 billion, increased by 13,50%), chemical products ($556.47 million, increased by 4.89%) and raw materials, textiles, garments, leathers and shoes ($402.17 million, decreased by 2.58%). The items with the highest growth rates in 2019 are cameras, cine cameras and components (reaching $6.44 million, increased by 114.93%), types of completely built-up cars ($135.83 million, increased by 74.64%), other conventional metal products ($15.98 million, increased by 73.64%), types of paper ($77.80 million, increased by 41.94%), gemstones, precious metals and related products ($78.48 million, increased by 37.28%) and computers, electronic products and components

Table 3 Some primary import goods from EU

	Goods	2017	2018	2019	2019/2018 (%)
01	Machinery and equipment	3.431,5	4.069,5	*3.909,9*	**−3,92**
02	Pharmaceuticals	1.440,3	1.438,8	*1.633,1*	**+13,50**
03	NPL textiles, leathers	312,6	412,8	*402,2*	**−2,58**
04	Metals	74,1	148,1	*174,0*	**+17,48**
05	Fertiliser	41,5	37,8	*29,4*	**−22,37**
06	Other transportation means	133,1	332,9	*257,1*	**−22,77**
07	Milk and dairy products	217,6	192,4	*214,9*	**+11,74**
08	Computers, phones	154,8	1.843,4	*2.514,4*	**+36,40**
09	Chemical products	221,3	530,5	*556,5*	**+4,89**
10	Car's components	512,1	248,2	*218,8*	**−11,85**
11	Completely built-up cars	115,3	77,8	*135,8*	**+74,64**

Unit USD million
Source The general department of Vietnam customs

($2.51 billion, increased by 36.40%). It is worth mentioning that the growth of some imported items decreased, such as iron and steel scrap ($59.69 million, decreased by 53.14%), ores and other minerals ($4.95 million, decreased by 29.17%), pesticides and raw materials ($81.16 million, decreased by 27.42%), chemicals ($195.56 million, decreased by 25.46%), other means of transport and spare parts ($257.16 million, decreased by 22.77%) and fertilizer of all kinds ($29.36 million, decreased by 22.37%).

Regarding Vietnam—EU trade relations

On 2018 March 26, The European Commission (EC) issued a decision to investigate trade remedies for 26 types of imported steels, including steels originated from Vietnam due to a sudden increase in steel imports. This move may lead to a rise in import taxes or imposition of quotas on specific types of Vietnamese steels; On 2018 June 26, the EC issued a decision to add two more types of steel products subject to investigation;

On 2018 June 23, the European Commission assessed that the situation of illegal, unreported and unregulated fishing in Vietnam had not made much progress since getting the "yellow card" (October 23, 2017). Exporting fishery products from Vietnam to Belgium and the EU still occurs normally due to the enormous demand for EU seafood imports, however the persistence of "yellow card" will more or less cause anxiety for both Vietnamese exporters and European importers. Vietnam fisheries management agencies and fishing vessel owners will have to increase management costs and invest in equipment to meet the IUU requirements;

On 2018 July 2, the European Commission issued a decision to strengthen the inspection of pesticide residues in some Vietnamese herbs and dragon fruits exported to the EU. This decision will increase the cost of testing and raise the risk of related products being rejected of clearance at EU ports.

Regarding Free Trade Agreement and Investment Protection Agreement between Vietnam and the European Union (EVFTA and EVIPA)

2019 June 30, under the witness of the Prime Minister Nguyen Xuan Phuc, Minister of Industry and Trade Tran Tuan Anh, representing Vietnam and European Commissioner for Trade Cecilia Malmstrom, Romanian Minister in charge of business, commerce and enterprise Stefan-Radu Oprea, representing the EU, officially signed the Vietnam—EU Free Trade Agreement (EVFTA).

The successful signing of this Agreement has marked a new milestone on the road of nearly 30 years of cooperation and development between Vietnam and the EU. That is also a positive message about Vietnam's determination to promote deep integration into the world economy in the context of complicated and unpredictable developments in the world's economic and political situation.

The strong commitment to open markets in the EVFTA Agreement will certainly promote Vietnam—EU trade relations, helping to expand the market for Vietnam's exports further. With the commitment to eliminating imports taxes up to nearly 100% of the tariff and trade value agreed by both sides, the opportunity to increase exports for Vietnam's advantageous products such as textiles, footwear, agriculture and fisheries (including rice, sugar, honey, vegetables), furniture, etc. is very significant; at the same time, it also enables Vietnamese consumers to access the supply of high-quality products and services from the EU in such areas as pharmaceuticals, health care, infrastructure construction, and public transport, etc.

Along with strengthening the overall relationship with the EU, the free trade agreement between Vietnam and the EU also creates excellent conditions for Vietnam and each member country to open up new cooperation opportunities based on advantages of each country, step-by-step bringing bilateral co-operations between Vietnam and each member country more and more into the quality and sustainability.

However, both sides still have to go one step further to put 2 agreements into implementation, submitting to the National Assembly of Vietnam and the European Parliament for ratification of the two agreements. For Vietnam, the Ministry of Industry and Trade will be responsible for preparing the EVFTA approval documents, the Ministry of Planning and Investment will be responsible for the EVIPA approval documents. The process of ratification of the Agreement shall comply with the procedure specified in the Treaty Law. The Government shall submit the dossier of application for approval to the President, and the President will decide to submit to the National Assembly for approval. For the EU, the approval process differs between EVFTA and EVIPA. Specifically, with EVFTA, only the European Parliament's approval is needed to come into force immediately. The EU called it "provisional" effect because afterward, in principle, EVFTA must still be ratified by the Parliament of 28 EU member countries. EVIPA is different, the European Parliament and the Parliament of all 28 member countries must ratify this Agreement for the Agreement to come into effect.

On 2020 January 21, the International Trade Commission (INTA) voted to ratify the two agreements under which the EVFTA Agreement received 29 votes in favor and EVIPA received 26 votes in favor. It is the highest vote rate compared to some recent FTAs between the EU and its partners. The European Parliament (EP) is expected to vote in the plenary session on 2020 February 12. After that, the two Agreements will need the National Assembly of Vietnam to approve (scheduled for the May 2020 meeting) to come into effect officially.

Vietnam—EU Investment Relations

EU Investment in Vietnam

In 2019, the EU had 2375 projects (182 projects more than 2018) from 27/28 EU countries that are still valid in Vietnam with a total registered investment capital of US$ 25.49 billion (an increase of $1.19 billion) accounting for 7.70% of number projects and 7.03% of the total registered investment capital of Vietnam. In particular, the Netherlands ranked first with 344 projects and $10.05 billion, accounting for 39.43% of total EU investment in Vietnam (increased by 26 projects and $692.76 million of investment capital). The United Kingdom ranked second with 380 projects and $3.72 billion of total investment capital, accounting for 14.58% of total investment capital (increased by 29 projects and $210.10 million of investment capital). France ranked third with 563 projects and total investment capital of $3.60 billion, accounting for 14.13% of total investment capital (increased by 23 projects but decreasing $72.07 million of investment capital).

In general, the European investors have the technology advantage, so they have positive contributions to create some emerging industries and products with high technological content. Some large EU corporations are operating effectively in Vietnam such as BP (UK), Shell Group (Netherlands), Total Elf Fina (France Belgium), Daimler Chrysler (Germany), Siemen, Alcatel Comvik (Sweden) and etc. The investment trend of the EU is mainly focused on the high-tech industries, however, there has been a tendency in recent years to centres more on service sectors (post and telecommunications, finance and office for retail).

The Vietnamese Investment in the EU

Regarding the investment of Vietnamese enterprises in the EU, it is not much in general and mainly focuses on some countries such the Netherlands, Czech and Germany. As of the end of 2018, Vietnam has 78 investment projects with a total registered capital of about 320.20 million USD in 10 EU countries (including: England, Poland, Belgium, Portugal, Germany, Netherlands, France, Czech Republic, Spain and Slovakia). In particular, these projects mainly came to Germany with 29 projects worth 120.3 million USD, England and British Virgin Islands with 20 projects worth 144.5 million USD, France with 10 projects worth 5.4 million USD and Slovakia with 2 projects worth 36.4 million USD.

(From the European-American Market Department, the Ministry of Industry and Trade of Vietnam).

2 EVFTA Review

The structure of EVFTA is significantly different from other previous free trade agreements that Vietnam has signed. Previous Free Trade Agreements had only average standards and mainly focused on tariff reduction, opening up services market but not exceeding commitments in the World Trade Organization (WTO).

With EVFTA, the Agreement commits to open the market up to more than 99% of tariff lines and trade turnover, the 0% tax rate will be applied to export items that the two sides can be strong with such as Vietnam's textiles, footwear, sandals, seafood, tropical agricultural products, furniture, etc. and Europe's automobile, machinery, equipment, alcohol, pharmaceuticals, temperate agricultural products.

Regarding trade in services, the commitments of both sides go beyond the commitments in the WTO framework. EU businesses will enjoy more incentives when investing and doing business in Vietnam, especially in areas where EU businesses can be strong such as banking, financial services, distribution, and transportation.

In addition to the above issues, there are also issues that Vietnam has never committed to such as: investment (both in production and services), policy for state-owned enterprises, public procurement, labor, the environment. Accordingly, the Agreement aims to ensure equality between all businesses, all fields, and puts state enterprises in the game of equal competition. State-owned enterprises currently operating in monopolistic sectors, when entering a competitive business, are also governed by this commitment. The coverage of the commitment is valid for both central SOE and local SOEs.

In general, in the context of Vietnam's extensive economic integration, the resonant impact of the EVFTA and signed FTAs is great, contributing to the economic development and innovation of Vietnamese enterprises and helping grow foreign and Vietnamese investors increasingly in FTA markets. For both Vietnam and the EU, it can be said that trade in goods will be the most promising sector at present as the elimination of tariffs will continue to boost the import and export activities of goods that have been and are developing strongly between two parties. For the EU, investment in a number of strong services such as banking, finance, distribution, and transportation will benefit as soon as the agreement comes into effect. In which:

2.1 In Main Terms of EU

Eliminating Customs Duties

The FTA will eliminate nearly all tariffs (over 99%): Vietnam will liberalize 65% of import duties on EU exports to Vietnam at entry into force, with the remainder of duties being gradually eliminated over a 10 year period. EU duties will be eliminated over a 7 year period.

A few examples:

Almost all EU exports of machinery and appliances will be fully liberalized at entry into force and the rest after 5 years.

Motorcycles with engines larger than 150 cc will be liberalized after 7 years and cars after 10 years, except those with large engines (>3000 cc for petrol, >2500 cc for diesel) which will be liberalized one year earlier.

The totality of EU textile fabric exports will be liberalized at entry into force.

Close to 70% of EU chemicals export will be duty free at entry into force and the rest after 3, 5 and 7 years.

Vietnam will also open its market for most EU food products, both primary and processed:

- Wines and spirits will be liberalized after 7 years.
- Frozen pork meat will be duty free after 7 years, beef after 3 years, dairy products after a maximum of 5 years, food preparations after a maximum of 7 years.
- Chicken will be fully liberalized after 10 years.

The EU will also eliminate duties with longer staging periods (up to 7 years) for some sensitive products, especially in the textile apparel and footwear sectors. To benefit from the preferential access, the strict rules of origin for garments will require the use of fabrics produced in Vietnam, with the only exception being of fabrics produced in South Korea, another FTA partner of the EU.

Only some sensitive agricultural products will not be fully liberalized, but the EU has offered access to Vietnamese exports via tariff rate quotas (TRQs): rice, sweet corn, garlic, mushrooms, sugar and high-sugar-containing products, manioc starch, surimi and canned tuna.

Reducing Non-tariff Barriers to European Exports

The EU and Vietnam have agreed to strengthen the disciplines of the WTO Technical Barriers to Trade (TBT) agreement. In particular, Vietnam has committed to increasing the use of international standards in drafting its regulations. The agreement also contains a chapter addressing Sanitary and Phytosanitary measures (SPS), specifically aimed at facilitating trade in plant and animal products, where the parties agreed on some important principles such as regionalization and the recognition of the EU as a single entity. These provisions will facilitate access for EU companies producing a large variety of products, including electrical appliances, IT, and food and drinks to the Vietnamese market.

Protecting European Geographical Indications

Farmers and small businesses producing food with traditional methods will benefit from the recognition and protection on the Vietnamese market—at a comparable level to that of EU legislation—of 169 European food and drink products from a specific geographical origin. This means that the use of Geographical indications (GIs) such as Champagne, Parmigiano Reggiano cheese, Rioja wine, Roquefort cheese or Scotch Whisky will be reserved in Vietnam for products imported from the European regions where they traditionally come from.

Vietnamese GIs too will be recognized as such in the EU, providing the adequate framework for further promoting imports of quality products such as Mộc Châu tea or Buôn Ma Thuột coffee. The agreement will allow new GIs to be added in the future.

Allowing EU Companies to Bid for Vietnamese Public Contracts

With this agreement EU companies will be able to bid for public contracts with, inter alia, Vietnamese ministries, including for infrastructure such as roads and ports, important state-owned enterprises such as the power distribution company and the nationwide railway operator, public hospitals and the two biggest Vietnamese cities, Hanoi and Ho Chi Minh City.

Creating a Level Playing Field for EU Companies and Innovative Products

With the disciplines agreed on State Owned Enterprises (SOEs) and subsidies, the EU-Vietnam FTA will level the playing field between SOEs and private enterprises when SOEs are engaged in commercial activities. There will also be rules on transparency and consultations on domestic subsidies. These are the most ambitious disciplines that Vietnam has ever agreed to.

On Intellectual Property Rights, Vietnam has committed to a high level of protection, going beyond the standards of WTO TRIPs agreement. With this agreement, EU innovations, artworks and brands will be better protected against being unlawfully copied, including through stronger enforcement provisions.

Opening the Vietnamese Market for EU Services Operators

Vietnam has committed to substantially improve access for EU companies to a broad range of services sectors, including business services, environmental services, postal and courier services, banking, insurance and maritime transport.

Promoting and Protecting Investment

Vietnam has committed to open up to investments in manufacturing in a number of key sectors, such as food products and beverages, fertilizers and nitrogen composites, tyres and tubes, gloves and plastic products, ceramics and construction materials.

On investment protection, both sides have already achieved a lot, including an agreement on key provisions on protection such as National Treatment and understandings on the main substantive investment protection rules. A permanent investment dispute resolution mechanism will be set up by creating an independent Investment Tribunal System.

Mechanism to Resolve Future Disagreements

The FTA creates a framework to resolve any future disagreements that may occur between EU and Vietnam about the interpretation and implementation of the agreement. It applies to most areas of the agreement and it is faster and more efficient in many aspects than the dispute settlement mechanism in the WTO.

Safeguarding Social and Environmental Protection Standards

The EU and Vietnam have agreed on a robust and comprehensive chapter on trade and sustainable development, with an extensive list of commitments, including:

Commitment to the effective implementation by each Party of the ILO core labor standards, ratified ILO Conventions (not only the fundamental ones) and ratified Multilateral Environmental Agreements and to ratification of not yet ratified fundamental ILO Conventions.

Promotion of Corporate Social Responsibility, including references to international instruments in this regard.

A dedicated article on climate change and commitments to the conservation and sustainable management of biodiversity (including wildlife), forestry (including illegal logging), and fisheries.

Promoting Democracy and Respect for Human Rights

In the preamble of the FTA the Parties reaffirm their commitment to the Charter of the United Nations signed in San Francisco on 26 June 1945 and have regard to the principles articulated in The Universal Declaration of Human Rights adopted by the General Assembly of the United Nations on 10 December 1948.

The FTA will also contain a legally binding link with the EU-Vietnam Partnership and Cooperation Agreement (PCA), signed in June 2012, which includes a human rights clause and provisions on cooperation on human rights.

The EU-Vietnam Partnership and Cooperation Agreement (PCA)

The PCA governs the overall relationship between the EU and Vietnam. It is based on shared interests and principles such as equality, mutual respect, the rule of law and human rights. It broadens the scope of the cooperation in areas such as trade, the environment, energy, science & technology, good governance, as well as tourism, culture, migration, counter terrorism and the fight against corruption & organized crime. Additionally, it allows Vietnam and the EU to further enhance cooperation on global and regional challenges, including climate change, terrorism and non-proliferation of weapons of mass destruction.

The PCA provides that human rights, democracy, and the rule of law are 'essential elements' in the overall relationship between the EU and Vietnam. Therefore, the link between the FTA and the PCA is important to ensure that human rights are also part of the trade relationship between the Parties.

2.2   In Main Terms of Vietnam

Recently ratified by Vietnam's National Assembly, the European Union Vietnam Free Trade Agreement (EVFTA) presents exciting opportunities in a multilateral trading partnership between both parties.

Not only has the agreement slashed tariffs on nearly 99 percent of all Vietnamese exports to the EU, but measures have also been taken to ensure that the FTA stays updated in the face of future agreements on both sides. For those seeking to tap into the agreement's cost-saving measures, an important consideration, and the focal point of this article, are the EVFTA's rules of origin.

In the face of low-cost sourcing options in close proximity to the Vietnamese market, and increasing regional integration on the part of the Association of Southeast Asian Nations (ASEAN)—of which Vietnam is a member, existing producers, and newcomers to the region alike will be hard-pressed to deny the benefits of regionally integrated supply chains.

However, while sourcing materials from third party states may decrease the overall cost of producing a given good, the introduction of third party inputs may compromise coverage under the EVFTA—leading to reduced competitiveness upon export to European markets.

The following paragraphs highlight the conditions under which goods will be granted coverage under EVFTA and outline the documentation that should be expected as part of the customs compliance process.

Qualifying for Tariff Reductions

Products will benefit from the tariff preferences under the EVFTA rules of origin provided that they can prove that they are "originating". Products are considered originating under the agreement if they meet one of the following requirements:

- Wholly obtained in Vietnam; and
- Products produced in Vietnam incorporating materials which have not been wholly obtained there, provided that such materials have undergone sufficient working or processing within Vietnam.
- While raw materials from Vietnam and goods produced in Vietnam using Vietnamese inputs easily fall into the wholly obtained category, many goods contain materials or components imported from countries not party to a trade agreement.

These goods must prove that the inputs that have been inputted have undergone specific levels of alteration within Vietnamese borders to tap into the benefits of the EVFTA. Many goods have set procedures that must be completed within Vietnam for the good in question to be considered originating. The chart below provides an example of the layout of product-specific working requirements found within the EVFTA text (Table 4).

Specific Products Under EVFTA

In addition to these procedures, certain areas of working are specifically noted for their inability to qualify as goods for originating status. These exemptions include the following:

- preserving operations to ensure that the products remain in good condition during transport and storage;
- breaking-up and assembly of packages;

Table 4 Sample of product specific working requirements

Layout of product specific working requirement		
HS heading	Description of product	Working or processing, carried out on non-originating materials, which confers originating status
Ex chapter 23	Residues and waste from the food industries; prepared animal fooder; except for:	Manufacture from materials of any heading, except that of the product
2030 and ex 2303	Residues of starch manufacture	Manufacture from material of any heading, except that of the product, in which the weight of the materials of chapter 10 used does not exceed 20% of the weight of the final product

- washing, cleaning; removal of dust, oxide, oil, paint or other coverings;
- ironing or pressing of textiles and textile articles;
- simple painting and polishing operations;
- husking and partial or total milling of rice;
- operations to color or flavor sugar or form sugar lumps;
- peeling, stoning, and shelling of fruits, nuts, and vegetables;
- sharpening, simple grinding or simple cutting;
- sifting, screening, sorting, classifying, grading, matching (including the making-up of sets of articles);
- simple placing in bottles, cans, flasks, bags, cases, boxes, fixing on cards or boards, and all other simple packaging operations;
- affixing or printing marks, labels, logos, and others like distinguishing signs on products or their packaging;
- simple mixing of products, whether or not of different kinds; mixing of sugar with any material;
- simple addition of water or dilution or dehydration or denaturation of products;
- simple assembly of parts of articles to constitute a complete article or disassembly of products into parts; and
- slaughter of animals.

EVFTA Compliance

Under the EVFTA, all firms exporting goods from Vietnam to the EU have two options concerning compliance. Depending on their status with the Vietnamese government, exporters must either complete a certificate of origin and origin declaration form or a specialized origin declaration form. Compliance associated with these alternatives is outlined below:

(a) *Certificate of Origin*
     EU based exporters should compile information available of the EVFTA to obtain a certificate of origin. Those seeking to export to the EU from Vietnam

will have to meet requirements that, while similar to those mentioned above, are outlined in Vietnamese legislation.

Vietnam's Ministry of Industry and Trade issued Circular No. 11/2020/TT-BTC implementing the rules of origin in the EVFTA. Vietnam's goods exported to the EU market will be granted the certificate of origin (C/O) form EUR 1 to enjoy preferential tariffs under the EVFTA. The circular will take effect on August 1, 2020.

In addition to the application forms listed above, it may be necessary to produce any of the following supporting information:

- direct evidence of the manufacturing or other processes carried out by the exporter or supplier to obtain the goods concerned, contained for example, in his accounts or internal book-keeping
- documents proving the originating status of materials used, issued or made out in a party, where these documents are used in accordance with domestic law
- documents proving the working or processing of materials in a party, issued or made out in a party, where these documents are used in accordance with domestic law
- proof of origin proving the originating status of materials used, issued or made out in a party in accordance with this protocol

(b) *Origin declaration*

Made out by any exporter for consignments the total value of which is to be determined in the national legislation of Vietnam and will not exceed US$6,600 (EUR 6000).

Exporters that have been approved by the Vietnamese government may forgo the issuance of a certificate of origin once their approval status has been relayed to relevant EU authorities. Instead, an Origin Declaration will be required upon export.

3 Opportunities and Challenges of Vietnam

3.1 *Opportunities*

- With a wide range of commitments and a high level of commitment, EVFTA will help Vietnam continue to reform its institutions and legal frameworks, and improve the business environment. Accordingly, EVFTA not only includes commitments related to trade liberalization and facilitation, but also includes commitments to liberalize trade in services and e-commerce, and government procurement, competition policy, state-owned enterprises, intellectual property, transparency, dispute settlement, etc. These commitments will be the driving force for Vietnam to reform its institutions and legal framework, improve business environment, creating favorable and safer investment conditions for investors.

– EVFTA offers great opportunities for trade development between Vietnam and the EU. Participation in EVFTA helps strengthen bilateral relations between Vietnam and the EU, contributes to diversifying markets and export products, in line with Vietnam's multilateral foreign policy in the international arena.
– EVFTA offers an opportunity to increase exports for advantageous Vietnamese products such as textiles, footwear, agricultural and fishery products, furniture, and mechanical engineering. Accordingly, the import tax for coffee, freshly processed fruits and vegetables, fresh fruit juices and flowers will be completely eliminated as soon as the agreement comes into effect. For seafood, 50% of tariff lines will be immediately eliminated, the remaining 50% will be deleted according to a schedule of 3–7 years. Particularly canned tuna and fish balls will be applied with tariff quotas.
– For rice, at the first time, a tariff rate quota will be applied with 80,000 tons enjoying a 0% tax rate, then the tax will be completely eliminated in 3–5 years. Tariffs on timber and wood products were reduced by as much as 83% over three to five years.
– For computers, electronic products and components, 74% of tariff lines will be eliminated import tax as soon as the agreement comes into effect. The remaining products will be exempt from import tax according to the 3–5 year roadmap.
– Textile and garment and footwear respectively 42.5 and 37% of tariff lines will be eliminated import tax as soon as the agreement comes into effect. The remainder will be deleted within 3–7 years.
– The mechanical engineering and manufacturing industry with 100% of tax lines will be exempt from import tax as soon as the Agreement comes into effect, from the average base tax of 1.8%.
– Implementation of commitments on technical barriers (TBT), sanitary and epidemiological measures (SPS), and rules of origin in EVFTA will help increase the added value of exports. When the Agreement is implemented, Vietnam will have to improve the quality of export products such as textiles, footwear, aquatic products, agricultural products, etc. according to technical standards, hygiene and export rules, the origin of products set out in the terms of the EVFTA. This is an opportunity to help Vietnam increase the value added in exports, restrict the export of raw products that do not meet international quality standards.
– The implementation of the FTA with the EU will contribute to the completion and development of the business environment, attracting direct investment from EU countries. According to the Foreign Investment Department, the Ministry of Planning and Investment of Vietnam, accumulated to April 2019, the EU is a major investment partner in Vietnam with valid 2244 projects, registered capital of 24, 67 billion VND, equivalent to 7.6% of the total FDI into Vietnam. The EU has invested in 18/21 industries according to the national economic analysis system, which concentrated in the field of manufacturing technology (accounting for 36.3% of total investment capital, mainly in industries, such as filtering petro-chemicals 11%, textiles 6.94%, electronics 6.4%, processing products 5.6%, cars and means of transport 5.2%); electricity and gas production and distribution 20.7%, real estate 11%, information and communication 6.6%. Therefore, FDI

from EU close donations to the database moved in the direction of activeness in Viet Nam.

– The EVFTA contributes to improving the quality of FDI inflows into Vietnam when the EU is the dominant partner in source technology, high technology and clean technology. FDI from the EU not only adds investment capital, but also can help Vietnam approach and catch up with the world's new development trends, thereby promoting a more human and sustainable economic development.

– However, EVFTA is only a supporting factor, not decisive for investment activities. To improve the quality of FDI inflows, Vietnam needs to continue to improve the business environment, improve the quality of human resources and technological level.

– In addition to the amount of FDI attracted to the sectors of industrial production. The Vietnam Nam-EU FTA also boosts EU FDI inflows into developing high-quality service sectors in Vietnam such as financial services, insurance, energy, telecommunications, ports and marine transportation. The reduction of conditions for EU service providers will make these industries more fiercely competition.

– Create opportunities for transferring new and modern technology. When implementing EVFTA, Vietnam will have access to modern and advanced technical technologies from EU countries that will promote production and export development, improve product quality, and increase labor productivity.

– Impact on labor structure shift in the agricultural sector to work in industry and services, increase labor productivity, reduce unemployment, increase income and living standards of the people.

– Improving the quality of human resources due to access to international principles and standards and commitments in EVFTA will motivate the training of a skilled and highly qualified workforce.

– Participating in EVFTA will help Vietnam take further steps in environmental protection. Along with rapid industrialization and urbanization, like many other developing countries, Vietnam is facing climate change and environmental pollution with the main problems of land degradation, forest degradation, biodiversity loss, water pollution, air pollution and solid waste management. Chapter 13 of this Agreement on Trade and Sustainable Development imposes obligations on both sides to enforce existing environmental standards, while also attracts trade or investment, and complies with all approved multilateral environmental agreements, such as those on climate change, endangered species, and biodiversity.

– EVFTA will help Vietnam improve the field of intellectual property protection, for the benefit of owners and consumers. When signing EVFTA, Vietnam will join the Internet Treaty of the World Intellectual Property Organization (WIPO). These Treaties set standards to prevent unauthorized online access or use of innovative products, to protect owners' rights, and to address challenges that technologies and methods of communication, new sets for intellectual property rights. Of course, protecting Intellectual Property Rights not only requires a comprehensive legal framework, it requires strong enforcement.

- Regarding imports: Thanks to Vietnam's import tax reduction schedule under the EVFTA, Vietnamese businesses will also benefit from imported goods and raw materials with good quality and stability at reasonable price from the EU. In particular, businesses will have the opportunity to access sources of machinery, equipment and high technology from EU countries, thereby improving productivity and improving the quality of their products. At the same time, goods and services imported from the EU into Vietnam will put pressure on domestic businesses to improve their competitiveness.

3.2 Challenges

- One of the biggest challenges facing Vietnam to make full use of EVFTA once the Agreement comes into effect is to rapidly improve production processes and in-production quality control to meet stringent import requirements into the EU market, as well as conditions for preferential tariff treatment of EVFTA.
- Especially for agricultural products, non-tariff barriers include: hygiene, quarantine, packaging, traceability and strict customs procedures, EU's standards are often the highest in the world which is a challenge for developing countries, including Vietnam.

The recent case in Vietnam's fisheries sector is in point. Seafood is one of the main export items and a strength of Vietnam in the international market. However, in October 2017, Vietnam had to receive a "yellow card" warning from the European Commission (EC) about illegal, unreported and unregulated fishing (IUU). This has a great impact on the reputation of Vietnamese seafood in the international market. The government, the fishing community, and fisheries businesses have taken action to tackle illegal fishing. However, Vietnam has a long coastline of 3260 km (excluding islands) with about 111,000 fishing boats (according to the Ministry of Agriculture and Rural Development) with mostly small boats, to completely solve the problem of exploitation for unauthorized seafood takes a long and laborious process.

- The issue of rules of origin (especially for textiles and garments) is one of the challenges facing Vietnam when signing and implementing an FTA which has been included in EVFTA. The reason is that Vietnam's textile production today is mainly product processing, most of the input value such as textile and garment raw materials is not originated from Vietnam. The requirement for rules of origin for fabric or yarn products to be manufactured in Vietnam will be a difficult condition for our country and the textile and garment industry will risk disadvantages when implementing the Agreement. However, in terms of origin for fishery and agricultural products, Vietnam can fully satisfy the requirements for these products.

– The huge competitive pressure in the domestic market due to the tariff cuts makes imported goods highly competitive. Goods imported from the EU to Vietnam will be easier and cheaper because they are not subject to import tax, and will have an advantage over domestic goods. If Vietnamese goods do not improve their quality and have price competition, it will be difficult to compete with goods imported from the EU when implementing the FTA.
– Domestic industries that are subject to fierce competition in the domestic market are mainly agricultural sectors such as livestock. Import tax on frozen pork will be 0% after 7 years; import tax on other types of meat will be 0% after 9 years; chicken meat will be exempt from import tax after 10 years; import duties will be eliminated after 3 years. This is a fairly long tariff elimination roadmap, so the impact of EVFTA on Vietnam's livestock industry will not be too sudden, and the industry will have a relatively long time to adjust and adapt to competition. Imports of meat products from the EU into Vietnam are still relatively small, however, when the current very high MFN tariffs (10–40%) are gradually reduced and eliminated at the end of the roadmap. The proportion and value of imports from the EU will increase significantly. Along with the tax cut, factors such as consumer sentiment (wanting to use hygienic and quality products) will significantly increase competition for Vietnam's livestock industry in the domestic market. This is the pressure on businesses in the livestock and food processing industries when the tax reduction roadmap is completed.

For the dairy industry, the entry into force of the EVFTA will put a competitive pressure on Vietnam's dairy enterprises because they have to compete with imported dairy products from the EU that have advantages in quality, nutrition and safety for the health of consumers. However, this pressure is currently negligible, due to the fact that Vietnam currently imports too much dairy products from the EU include: whey milk and variations, butter, cheese, powdered milk and powdered cream. These products are accounting for a small proportion of dairy product consumption in the Vietnamese market.

– According to EVFTA, there are strict regulations set for Vietnam on anti-dumping, subsidy and use of trade defense tools. For some industries with export advantages, the EU will require a reduction of non-tariff barriers that Vietnam is currently applying as subsidies from the Government. In addition, the introduction of requirements for non-tariff measures will affect some major Vietnamese exports such as textiles, footwear, and seafood if Vietnam does not request in line with its interests.
– Vietnam needs to adapt to resolving trade disputes such as labor and trade unions. EVFTA has a chapter on trade and sustainable development that includes commitments from both sides to labor and environmental standards. The Agreement also establishes institutions for the monitoring and enforcement of commitments

between the parties. In the process of EVFTA implementation, when problems related to labor and trade union occur, it is necessary to deal with the common regulations between the two sides, this is also a challenge for Vietnam because there is not yet a problem.

- Implementation of commitments must comply with the prescribed roadmap. In order to comply with the roadmap, Vietnam requires a great effort in the production and export process and fully comply with the commercial commitments between the two sides.
- The lack of qualified and skilled human resources that can meet the change and expansion in the production process applying modern high technologies when implementing EVFTA will be one of the big challenges for Vietnam. Therefore, Vietnam needs to prepare a highly qualified workforce that effectively allocates resources across all sectors, from infrastructure to finance to serve the digital economy and increase efficiency of activities of key economic sectors to serve the role of promoting development.
- The legal environment affects trade, with an incomplete legal environment that will hinder the attraction of EU businesses to invest in Vietnam. In addition, the infrastructure has not met the needs of production expansion and development.
- One of the issues that EU investors are concerned about, especially in the fields related to technology, is the protection of intellectual property rights. Statistics show that most Vietnamese enterprises are not yet interested in intellectual property, while this is the top EU requirement for goods entering this market. Therefore, in order to benefit from the Agreement, Vietnam needs to pay special attention to the intellectual property rules in the EVFTA. To meet the intellectual property rules in EVFTA, Vietnam needs to review the Intellectual Property Law to align with deeper commitments in EVFTA and other new generation FTAs, as well as ensure real fully and seriously test commitments to create confidence for investors.
- Budget source will decrease: When tax is reduced to 0% according to EVFTA, tax revenue will decrease due to lower import tax on goods from EU to Vietnam. Considering the above, in the long term, the reduction of import tax will have a positive impact on the production and consumption development of Vietnam.

In general, opportunities and challenges always go hand-in-hand, Vietnam needs to make good use of opportunities and respond promptly to challenges encountered in the implementation of the Vietnam-EU Free Trade Agreement. If you do not make good and effective use of opportunities, it will turn into a challenge, and vice versa, if you know how to properly handle and deal with challenges, the challenges you pose will turn into opportunities to promote commercial development and international trade of Vietnam through EVFTA.

4 Solutions for Vietnam Products to EU Market

4.1 On the Businesses' Side

– Improve the competitiveness of enterprises and export products:

When participating in EVFTA, businesses need to improve their competitiveness in the following ways:

(a) Diversifying export products because EU countries are potential markets with rich consumer demand for many different types of goods;
(b) To research export products in terms of price and quality. Special attention should be paid to the design of products, so businesses must understand consumer needs, customs and culture of EU export markets.
(c) Innovating technology, management skills to improve productivity and product quality;
(d) Invest in branding for the product to create a long-term foothold for the business.

– Promote trade promotion in EU countries' markets

To penetrate and expand market share in EU markets, businesses need to focus on and promote trade promotion activities. Enterprises need to learn about market characteristics such as laws, consumer tastes, how to do business, organize market research, participate in international fairs and exhibitions in export markets.

– Strengthen cooperation between businesses

Enhance the role of industry associations in linking businesses. The connection between businesses will help businesses be stronger in protecting their interests when there are disputes and lawsuits in trade. Cooperation between enterprises is an important measure to limit the negative aspects of the competition mechanism. In addition, businesses need to exchange business information, associate in opening bonded warehouses, opening trade centers, and coordinating in investment projects.

– Prepare and respond promptly to commercial disputes

In implementing EVFTA, opening an extensive book market is more prone to commercial disputes. Therefore, businesses need to deal well with commercial disputes, such as anti-dumping lawsuits; need to master the specific laws and regulations in the EU market and coordinate with industry associations in resolving trade disputes.

Enterprises need to maintain a transparent and scientific accounting system in order to fully and accurately provide the necessary data in the process of investigating and handling commercial disputes.

– Developing human resources in enterprises

Human resources are an important factor in EVFTA implementation, in order to meet the requirements and commitments in EVFTA, a qualified human resource is required. Enterprises should have policies to develop highly qualified human resources.

4.2 On the Government's Side

– Continuing to review perfect institutions and policies for business investment, accordingly, the government will study, review and propose to the National Assembly to consider amending, supplementing or promulgating a number of important laws such as the Investment Law, the Enterprise Law, the Investment Law according to the Public Partnership Mode Law, Land Law, Construction Law, Environmental Protection Law, Labor Code and some tax laws, etc. to comply with the regulations and the principles of cooperation in the framework of EVFTA.
– Improving the business environment and protecting the legitimate interests of investors, joint commitments practical actions will be proportional to the confidence of investors.
– Actively signing mutual recognition agreements and equivalence agreements in each specific case with the EU to reduce costs of complying with standards of technical measures, sanitary measures and phyto-sanitary measures of the EU.
– Help businesses raise awareness in dealing with non-tariff barriers, or renegotiate with importers to help businesses overcome barriers.

5 Conclusion

EVFTA has opened up many opportunities as well as challenges for import and export activities of Vietnamese enterprises. Therefore, along with the improvement of the state's investment policy and environment, businesses need to actively seek information about the EVFTA to grasp Vietnam's commitments and EU commitments, especially, information on tax incentives for commodities that have potential to network, or for export in the near future. Finding and understanding the commitments, challenges, and opportunities related to the industry and sector will help the reposition, role, restructure markets, partners, and supply sources, attaching special importance to meet the requirements of the EU's rules of origin and technical standards. Moreover, enterprises actively respond to changes in the business environment by developing and adjusting business plans for the medium and long term in order to boost the flow of goods into the EU market. At the same time, it is necessary to ensure the quality of goods, meet the set standards, and build and develop new brands that can create competitiveness for Vietnamese goods.

References

1. Vietnam – EU Bilateral Relationship Article, accessed on 12 Feb 2020. Available at: http://evfta.moit.gov.vn/?page=overview&category_id=fb203c7b-54d6-4af7-85ca-c51f227881dd
2. Authors: Report of the EVFTA's effects to policy and institution. Central Institute for Economic Management (CIEM) (2017)
3. Nguyen T.B.: The EVFTA's Impacts To Vietnamese Economy. Finance News. Ministry of Finance, Vietnam (2016)
4. Introduction of EVFTA. http://evfta.moit.gov.vn
5. Nguyen, T.C., Pham, H.N.: The Impacts of EVFTA Towards Vietnam's Economy. Hanoi Socio-economic Development Research Institute (2017)

Index

© Institute of Technology PETRONAS Sdn Bhd 2022
S. A. Abdul Karim (eds.), *Shifting Economic, Financial and Banking Paradigm*,
Studies in Systems, Decision and Control 382,
https://doi.org/10.1007/978-3-030-79610-5